TUTOR DELIVERY PACK
BIOLOGY
— GCSE —
HIGHER

Author: Janette Gledhill

BIOLOGY HIGHER

Contents

Page		Specification links
3	How to use this pack	
5	Information for parents and guardians	
7	Specification guidance	
11	Mapping of lessons to AQA GCSE (9–1) Combined Science: Trilogy	
13	Needs analysis	
15	Progress report	
16	End-of-lesson report	
17	1 Diagnostic lesson	Maths skills; investigative skills

Cell biology

Page		Specification links
23	2 Cell structure and microscopes	4.1.1.1 Eukaryotes and prokaryotes; 4.1.1.2 Animal and plant cells; 4.1.1.5 Maths skills 1a, 1b, 2h
29	3 Cell specialisation and differentiation	4.1.1.3 Cell specialisation; 4.1.1.4 Cell differentiation
35	4 Culturing microorganisms	4.1.1.6 Culturing microorganisms; Maths skills 1b, 5c
41	5 Cell division	4.1.2.1 Chromosomes; 4.1.2.2 Mitosis and the cell cycle; 4.1.2.3 Stem cells
47	6 Transport in cells	4.1.3.1 Diffusion; 4.1.3.2 Osmosis; 4.1.3.3 Active transport; Maths skills 1c, 4c, 5c

Organisation

Page		Specification links
53	7 The digestive system and enzymes	4.2.1 Principles of organisation; 4.2.2.1 The human digestive system
59	8 The heart, blood vessels and blood	4.2.2.2 The heart and blood vessels; 4.2.2.3 Blood
65	9 Non-communicable diseases	4.2.2.4 Coronary heart disease; 4.2.2.5 Health issues; 4.2.2.6 The effect of lifestyle; 4.2.2.7 Cancer; Maths skills 4a
71	10 Plant tissues, organs and systems	4.2.3.1 Plant tissues; 4.2.3.2 Plant organ systems; Maths skills 2b

Infection and response

Page		Specification links
77	11 Communicable diseases – viral and bacterial diseases	4.3.1.1 Communicable (infectious) diseases; 4.3.1.2 Viral diseases; 4.3.1.3 Bacterial diseases
83	12 Communicable diseases – fungal and protist diseases	4.3.1.4 Fungal diseases; 4.3.1.5 Protist diseases
89	13 Defence systems, vaccination, antibiotics and painkillers	4.3.1.6 Human defence systems; 4.3.1.7 Vaccination; 4.3.1.8 Antibiotics and painkillers
95	14 Drug development and monoclonal antibodies	4.3.1.9 Discovery and development of drugs; 4.3.2.1 Producing monoclonal antibodies; 4.3.2.2. Uses of monoclonal antibodies
101	15 Plant diseases	4.3.3.1 Detection and identification of plant diseases; 4.3.3.2 Plant defence responses

Bioenergetics

Page		Specification links
107	16 Photosynthesis	4.4.1.1 Photosynthetic reaction; 4.4.1.2 Rate of photosynthesis; 4.4.1.3 Uses of glucose from photosynthesis; Maths skills 3d, 4a, 4d
113	17 Respiration	4.4.2.1 Aerobic and anaerobic respiration; 4.4.2.2 Response to exercise; 4.4.2.3 Metabolism

BIOLOGY HIGHER

Contents

Page		Specification links
	Homeostasis and response	
119	18 Homeostasis and the nervous system	4.5.1 Homeostasis; 4.5.2.1 Structure and function of the human nervous system; 4.5.2.4 Control of body temperature
125	19 The human brain and eye	4.5.2.2 The brain; 4.5.2.3 The eye
131	20 Hormonal coordination in humans	4.5.3.1 Human endocrine system; 4.5.3.2 Control of blood glucose concentration; Maths skills 1c
137	21 Water and nitrogen balance	4.5.3.3 Maintaining water and nitrogen balance in the body
143	22 Human reproduction and contraception	4.5.3.4 Hormones in human reproduction; 4.5.3.5 Contraception
149	23 Treating infertility and negative feedback	4.5.3.6 The use of hormones to treat infertility; 4.5.3.7 Negative feedback
155	24 Plant hormones	4.5.4.1 Control and coordination; 4.5.4.2 Use of plant hormones
	Inheritance, variation and evolution	
161	25 Types of reproduction and meiosis	4.6.1.1 Sexual and asexual reproduction, 4.6.1.2 Meiosis, 4.6.1.3 Advantages and disadvantages of sexual and asexual reproduction
167	26 DNA structure and the genome	4.6.1.4 DNA and the genome; 4.6.1.5 DNA structure
173	27 Inheritance and sex determination	4.6.1.6 Genetic inheritance; 4.6.1.7 Inherited disorders; 4.6.1.8 Sex determination; Maths skills 1c, 2e
179	28 Genetic understanding, variation and selective breeding	4.6.2.1 Variation; 4.6.2.3 Selective breeding; 4.6.3.3 The understanding of genetics; Maths skills 2c
185	29 Genetic engineering and cloning	4.6.2.4 Genetic engineering; 4.6.2.5 Cloning
191	30 Evolution and speciation	4.6.2.2 Evolution; 4.6.3.1 Theory of evolution; 4.6.3.2 Speciation
197	31 Evidence for evolution – extinction	4.6.3.4 Evidence for evolution; 4.6.3.5 Fossils; 4.6.3.6 Extinction; 4.6.3.7 Resistant bacteria
	Ecology	
203	32 Classification and communities	4.6.4 Classification of living organisms; 4.7.1.1 Communities
209	33 Adaptations, abiotic and biotic factors	4.7.1.2 Abiotic factors; 4.7.1.3 Biotic factors; 4.7.1.4 Adaptations; Maths skills 2d, 4a
215	34 Organisation, cycling of materials and decomposition	4.7.2.1 Levels of organisation; 4.7.2.2 How materials are cycled; 4.7.2.3 Decomposition; Maths skills 2b, 2d, 2f
221	35 Environmental change, biodiversity and waste management	4.7.2.4 Impact of environmental change; 4.7.3.1 Biodiversity; 4.7.3.2 Waste management
227	36 Land use, deforestation and global warming	4.7.3.3 Land use; 4.7.3.4 Deforestation; 4.7.3.5 Global warming; 4.7.3.6 Maintaining biodiversity
233	37 Trophic levels in an ecosystem	4.7.4.1 Trophic levels; 4.7.4.2 Pyramids of biomass; 4.7.4.3 Transfer of biomass; Maths skills 1c, 2c
239	38 Food production	4.7.5.1 Factors affecting food security; 4.7.5.2 Farming techniques; 4.7.5.3 Sustainable fisheries; 4.7.5.4 Role of biotechnology
245	Glossary	
249	*Revise* mapping guide	

BIOLOGY HIGHER

How to use this pack

The *Tutors' Guild* GCSE Biology Higher Tutor Delivery Pack gives you all of the tools you need to deliver effective Biology lessons to GCSE students who are sitting the Higher tier papers of AQA GCSE (9–1) Biology and AQA GCSE (9–1) Combined Science: Trilogy.

Lessons

There are 38 one-hour, six-page lessons in this Tutor Delivery Pack. Most tutors working for a full year will have around 38 lessons with a student. If you have less contact time, you can choose which lessons are most important to the student. Each lesson is standalone and can be taught independently from those preceding it.

Lesson plans

The first page of each lesson is your *lesson plan*. It is designed specifically for tutors and is intended to guide you through a one-hour session in either a one-to-one or small group setting. It is not designed to be student-facing.

Learning objectives and specification links

At the top of each lesson plan, you will find two lists. The first – *learning objectives* – is a list of your aims for the lesson. The learning objectives will be informed by the specification but may have been rephrased to make sure they are accessible to and useful for everyone. You can discuss these with the student or use them for your own reference when tracking progress. The second list – *specification links* – shows you where in the specification you can find the objectives relevant to the lesson. You can find out more about the specification on pages 7–10.

Activities

The first five minutes of your lesson should be spent reviewing the previous week's homework. You should not mark the homework during contact time: instead, use the time to talk through what the student learned and enjoyed, and any difficulties they encountered.

The final five minutes should be used to set homework for the forthcoming week. There are three ways to do this: using the *end-of-lesson report* on page 16; orally with a parent or guardian; or simply using the *homework activity sheet* on the fifth page of each lesson.

In each lesson plan, you will find four types of activities.
- *Starter activities* are 5–10 minutes each and provide an introduction to the topic.
- *Main activities* are up to 40 minutes long and are more involved, focussing on the main objectives of the lesson. Some lessons have more than two main activities. Where this is the case, only the first two lessons have associated activity sheets.
- *Plenary activities* are 5–10 minutes each, require little to no writing and recap the main learning points or prepare for the homework.
- *Homework activities* can be up to an hour long and put learning into practice.

In the lesson plan, you will find a page reference (where the activity is paper-based), a suggested timeframe and teaching notes for each activity. The teaching notes will help to guide you in delivering the activity and will also advise you on any common misconceptions associated with the topic.

Support and extension ideas

This pack is aimed at students who are targeting grades 4–9, but every student is different: some will struggle with activities that others working at the same level find straightforward. In these sections, you will find ideas for providing some differentiation throughout the activities.

BIOLOGY HIGHER

How to use this pack

Progress and observations

This section is left blank for you to use as appropriate. You can then use the notes you make to inform assessment and future lessons, as well as to inform progress reports to parents or guardians.

Activities

There are four student-facing activity sheets for each lesson: one for the starter activities; two for the main activities and one for the homework activity. On each sheet, you'll find activity-specific lesson objectives, an equipment list and a suggested timeframe. All activities are phrased for one-to-one tutoring but are equally as appropriate for small group settings. If you have a small group and the task asks you to work in pairs or challenge each other, ask the students to pair up while you observe and offer advice as necessary. Where appropriate, answers can be found on the sixth page of the lesson.

Diagnostics

The first lesson in this pack is a diagnostic lesson, designed to help you find out more about your student: their likes and dislikes; strengths and weaknesses and personality traits. As well as the diagnostic lesson, the needs analysis section (pages 13–14) allows you, the student and the student's parents or guardians to investigate together which areas of the subject will need greater focus. Together, these sections will help you deliver the most effective, best value tuition.

Progress report

This can be used to inform parents or guardians or for your own planning as frequently or infrequently as is useful for you. Spend some time discussing the statements on the report with the student. Be prepared, though – some students will tell you there isn't anything that they enjoy about the subject!

End-of-lesson report

Parent participation will vary greatly. The end-of-lesson report is useful for efficiently feeding back to parents or guardians who prefer an update after each lesson. There is space to review completed homework and achievements in the lesson, as well as space for the student to explain how confident they feel after the lesson. Finally, there is a section on what steps, including homework, the parent and student can take to consolidate learning or prepare for the following week. The end-of-lesson report may also be useful for communicating with some parents or guardians who speak English as a second language, as written information may be easier to follow.

Certificates

In the digital version of this pack, you will find two customisable certificates. These can be edited to celebrate achievements of any size.

BIOLOGY HIGHER

Information for parents and guardians

Introduction

Your child's tutor will often make use of resources from the *Tutors' Guild* series. These resources have been written especially for the new 9–1 GCSEs, and are tailored to the AQA GCSE (9–1) Biology and the AQA GCSE (9–1) Combined Science: Trilogy specifications. The tutor will use their expert knowledge and judgement to assess the student's current needs. This will allow them to target areas for improvement, build confidence levels and develop skills as quickly as possible to ensure the best chance of success.

Just as a classroom teacher might do, the tutor will use lesson plans and activities designed to prepare the student for the 9–1 GCSEs. Each set of resources has been designed by experts in GCSE Biology and reviewed by tutors to ensure it offers great quality, effective and engaging tuition.

Getting started

Before tuition can begin, the tutor will need to know more about your motives for employing them in order to set clear, achievable goals. They will also try to learn more about the student to ensure lessons are as useful and as engaging as possible.

To gather this information, the tutor will work through the *needs analysis* pages of this pack with you. It shouldn't take too long, but it will really maximise the value of the tuition time you pay for. You could also take this opportunity to discuss with the tutor any questions or concerns you may have.

Lessons and homework

Each lesson will have the same structure: there will be a starter, which is a quick introduction to the topic; some main activities, which will look at the topic in greater detail; and a plenary activity, which will be used to round off the topic. Throughout the year, the student will become increasingly confident with the content of the specification, but will also improve his or her speaking, writing, reading, listening and co-ordination skills through a carefully balanced range of activities.

At the end of each lesson, the tutor will set some homework, which should take no longer than an hour to complete. If you don't want the tutor to set homework, please let them know. If you are happy for homework to be given, they will either discuss the homework task with you at the end of the lesson or give you an end-of-lesson report. All of the homework activities are designed to be completed independently, but if you would like to help with completion of homework, the tutor will be able to tell you what you can do.

Further support

Parents and guardians often ask a tutor what else they can do to support their child's learning or what resources they can buy to provide extra revision and practice. As a Pearson resource, *Tutors' Guild* has been designed to complement the popular *Revise* series. Useful titles you may wish to purchase include:
- *Revise* AQA GCSE (9–1) Biology Higher Revision Guide (ISBN: 9781292135038)
- *Revise* AQA GCSE (9–1) Biology Higher Revision Workbook (ISBN: 9781292135014)
- *Revise* AQA GCSE (9–1) Combined Science: Trilogy Higher Revision Guide (ISBN: 9781292131627)
- *Revise* AQA GCSE (9–1) Combined Science: Trilogy Higher Revision Workbook (ISBN: 9781292131689)

Using pages 249–250 of this pack, the tutor will be able to tell you which pages of the *Revise* resources are appropriate for each lesson.

Information for parents and guardians

What's in the test?

You may have heard a lot about the new 9–1 GCSEs from your child's school, from other parents or in the media. Here is a breakdown of the AQA GCSE (9–1) Biology exam.

Students will sit two papers that assess learning against the AQA GCSE (9–1) Biology specification.

Paper 1: *Topics 1–4: Cell biology; Organisation; Infection and response; and Bioenergetics. (50% of the total marks)*
Students are given 1 hour and 45 minutes to complete Paper 1. It is made up of multiple-choice, structured, closed short answer and open response questions.

Paper 2: *Topics 5–7: Homeostasis and response; Inheritance, variation and evolution; and Ecology. (50% of the total marks)*
Students are given 1 hour and 45 minutes to complete Paper 2. Again, it is made up of multiple-choice, structured, closed short answer and open response questions.

AQA GCSE (9–1) Combined Science: Trilogy

This pack can also be used to teach AQA GCSE (9–1) Combined Science: Trilogy. A breakdown of the differences between the AQA GCSE (9–1) Biology and AQA GCSE (9–1) Combined Science: Trilogy specifications can be found on pages 11–12.

Here is a breakdown of the Biology section of the AQA GCSE (9–1) Combined Science exam.

Paper 1: *Topics 1–4: Cell biology; Organisation; Infection and response; and Bioenergetics. (16.7% of the total marks)*
Students are given 1 hour and 15 minutes to complete Paper 1. It is made up of multiple-choice, structured, closed short answer and open response questions.

Paper 2: *Topics 5–7: Homeostasis and response; Inheritance, variation and evolution; and Ecology. (16.7% of the total marks)*
Students are given 1 hour and 15 minutes to complete Paper 2. It, again, is made up of multiple-choice, structured, closed short answer and open response questions.

Results and grades

GCSE results day is typically the third or fourth Thursday in August. It is the same day across the country, so you can find out the exact date online. On results day, students will be given a slip of paper (or one per exam board, if the school hasn't collated them) with an overall grade for each GCSE. Grades for the 9–1 GCSE in Biology are given as numbers (9–1) instead of letters (A*–U). The diagram below shows roughly how the old-style grades translate to the new ones.

A*	A	B	C	D	E	F	G	U
9 8	7 6	5	4	3	2	1		U

As you can see, the new grade 9 is pitched higher than an A*. There is also a wider spread of grades available for students whose target would previously have been a B/C. GCSE (9–1) Biology is a tiered exam, with the Foundation papers aimed at those students targeting grades 1–5, and the Higher papers aimed at students hoping for grades 4–9.

BIOLOGY HIGHER

Specification guidance

The new AQA GCSE (9–1) Biology qualification was introduced for first teaching in 2016 and first assessment in 2018.

If you have experience in tutoring or teaching the previous curriculum, much of the content and assessment will be familiar. If this is the case, please turn to pages 8–10 for guidance on what has changed.

If you are new to tutoring GCSE Biology, this page will give you a brief introduction to the qualification before you move on to pages 8–10. Further guidance on particular areas of the specification – including common misconceptions and barriers to learning – can be found in the lesson plans throughout this book. Full details of the specification can be found on the AQA website.

Key facts

Subject content

There are seven areas of biology that will be assessed:
- Cell biology
- Organisation
- Infection and response
- Bioenergetics
- Homeostasis and response
- Inheritance, variation and evolution
- Ecology

The seven domains are further broken down into smaller content areas. These are set out from page 15 onwards of the AQA specification, which can be found on the AQA website.

Each lesson plan in this pack highlights which areas of the specification are covered. In order to maximise your student's chances of success, the pack covers the most important specification areas and those students struggle with the most; it is intended to supplement and enhance classroom teaching and does not therefore cover the entire specification.

Exam papers

Both Foundation and Higher students will sit two exam papers:

- **Paper 1**
  - Topics 1–4: Cell biology; Organisation; Infection and response; and Bioenergetics
  - 1 hour 45 minutes
  - 100 marks
  - 50% of GCSE
  - Foundation/Higher Tier
  - Multiple choice, structured, closed short answer and open response questions.
- **Paper 2**
  - Topics 5–7: Homeostasis and response; Inheritance, Variation and evolution; and Ecology
  - 1 hour 45 minutes
  - 100 marks
  - 50% of GCSE
  - Foundation/Higher Tier
  - Multiple choice, structured, closed short answer and open response questions.

As the breakdown above shows, each exam paper is given an equal weighting: half of the total available marks are available for each paper. Foundation students will be able to access grades 1–5 and Higher students will be able to access grades 4–9.

At least 15% of marks will be allocated to testing students' experience of practical science, and 10% of marks will be given for maths skills.

BIOLOGY HIGHER

Specification guidance

What's changed?

Key changes

There are several key changes to the AQA GCSE (9–1) Biology course, brought about by new Ofqual requirements. The hope is that increasing the demand of GCSE Biology will better prepare students to apply their learning in everyday life, in work and in further studies. The new course will be more demanding in the following ways.

- **The grading system has changed.**

  The A* to G grades will be replaced by 9 to 1 for GCSE Biology. Combined Science will have a 17-point grading scale, from 9–9, 9–8 through to 2–1, 1–1.

- **There is more subject content.**

  You'll have more topics to cover with your student, and the topics will be denser. This will change how you teach the course: will you recommend increased contact time, set more independent work or prioritise which content you cover?

- **The level of demand increases.**

  Questions become steadily more demanding within each topic area and throughout the paper, giving students of all abilities the chance to gain marks on each topic.

- **Total exam time has increased.**

  The minimum exam time for GCSE Biology is 3 hours 30 minutes. The minimum exam time for Combined Science is 7 hours.

- **There are no controlled assessments.**

  Controlled assessment will no longer be carried out, but will instead be replaced with exam questions testing students' understanding of scientific experimentation, including use of apparatus and techniques. Students are now expected to carry out a number of required practical activities in class, their knowledge of which will then be tested in the exam.

- **Quality of Written Communication (QWC) is no longer assessed.**

  Marks are no longer available for the quality of written communication. However, logical structuring of extended responses is still necessary to achieve a good grade in GCSE (9–1) Science qualifications.

New content

New topics include the following:

- Microscopy
- Health issues
- Cancer
- Viral, bacterial, fungal and protist diseases (Higher Tier only).

To make way for this new content, some previously covered content has been omitted from the new specification, including some aspects of the subject areas Cell structure, Diet and exercise and Genetic variation.

While this pack will help you to deliver the new content, you should make sure you are familiar and comfortable with the new topics and the best practices for teaching them. A full list of new content can be found on AQA's website.

BIOLOGY HIGHER

Specification guidance

Science and maths skills

The new GCSE (9–1) Biology qualification places special emphasis on mathematical skills, and on learning to work scientifically. You need to make sure you are able to help your students develop such skills.

Working scientifically

GCSE (9–1) Biology covers a range of scientific skills, including experimenting, observing and analysing results.

1. **Development of scientific thinking**

   This section covers a range of skills related to scientific thought, including: understanding scientific methods and theories; using scientific models; evaluating scientific risks; consideration of ethical issues.

2. **Experimental skills and strategies**

   This section covers a range of skills and strategies, including: developing hypotheses based on scientific theories; applying knowledge to select apparatus; sampling techniques; evaluation of scientific methods.

3. **Analysis and evaluation**

   This section covers skills relating to the collection, presentation and analysis of data, including: translating data from one form to another; mathematical and statistical analysis; interpreting observations; communicating scientific rationale.

4. **Scientific vocabulary, quantities, units, symbols and nomenclature**

   This section includes: using scientific vocabulary; using units correctly; using prefixes and powers of ten for orders of magnitude; using an appropriate number of significant figures.

The lesson plans in this pack indicate where key areas of working scientifically have been covered.

Mathematical skills

In GCSE (9–1) Biology, a minimum of 10% of marks will test the student's skills in mathematics. This is compared to 20% in GCSE Chemistry, and 30% in GCSE Physics.

Students will be required to demonstrate the following maths skills in GCSE (9–1) Biology:

- Arithmetic and numerical computation
- Handling data
- Algebra
- Graphs
- Geometry and trigonometry.

These five areas are described in full from page 83 of the AQA specification, which can be found on the AQA website.

In Foundation Tier papers maths skills will be tested up to Key Stage 3 standard. In Higher Tier papers maths skills requirements will not be lower than the requirements for the Foundation Tier in a GCSE qualification in mathematics.

The lesson plans in this pack indicate which areas of maths skills are covered in individual lessons.

BIOLOGY HIGHER

Specification guidance

Practical work

Students will carry out eight practicals for each of Biology, Chemistry and Physics, and 16 for Combined Science, which are all outlined in the relevant AQA specifications. This pack will not include practical work in the lessons, but will cover the relevant practical methods and apparatus needed for the exam.

- **Required practical activity 1**

  Use a light microscope to observe, draw and label a selection of plant and animal cells.

- **Required practical activity 2 (Biology only)**

  Investigate the effect of antiseptics or antibiotics on bacterial growth using agar plates and measuring zones of inhibition.

- **Required practical activity 3**

  Investigate the effect of a range of concentrations of salt or sugar solutions on the mass of plant tissue.

- **Required practical activity 4**

  Use qualitative reagents to test for a range of carbohydrates, lipids and proteins.

- **Required practical activity 5**

  Investigate the effect of pH on the rate of reaction of amylase enzyme.

- **Required practical activity 6**

  Investigate the effect of light intensity on the rate of photosynthesis using an aquatic organism such as pondweed.

- **Required practical activity 7**

  Plan and carry out an investigation into the effect of a factor on human reaction time.

- **Required practical activity 8 (Biology only)**

  Investigate the effect of light or gravity on the growth of newly germinated seedlings.

- **Required practical activity 9**

  Measure the population size of a common species in a habitat.

- **Required practical activity 10 (Biology only)**

  Investigate the effect of temperature on the rate of decay of fresh milk by measuring pH change.

AQA GCSE (9 -1) Combined Science: Trilogy

This pack can also be used to teach the AQA GCSE (9–1) Combined Science: Trilogy specification. This exam will cover the same seven areas of Biology as those taught in AQA GCSE (9–1) Biology, along with topics from the AQA GCSE (9–1) Chemistry and Physics specifications. The GCSE (9–1) Combined Science: Trilogy qualification is linear, meaning that students will sit all their exams at the end of the course.

Exam papers

For the Biology section of the exam, both Foundation and Higher students will sit two exam papers:

- **Biology Paper 1**
  - Topics 1–4: Cell biology; Organisation; Infection and response; and Bioenergetics
  - Written exam: 1 hour 15 minutes
  - 70 marks; 16.7% of GCSE
  - Foundation and Higher Tier
  - Multiple choice, structured, closed short answer and open response questions.
- **Biology Paper 2**
  - Topics 5–7: Homeostasis and response; Inheritance, Variation and evolution; and Ecology
  - Written exam: 1 hour 15 minutes
  - 70 marks; 16.7% of GCSE
  - Foundation and Higher Tier
  - Multiple choice, structured, closed short answer and open response questions.

A map for teaching AQA GCSE (9–1) Combined Science: Trilogy using this pack is available on pages 11–12. Please note that this pack is not intended to be used to teach AQA GCSE (9–1) Combined Science: Synergy.

Mapping of lessons to AQA GCSE (9–1) Combined Science: Trilogy

Lesson	Specification reference Combined	Biology	Suggested adaptation for Trilogy
1	N/A	N/A	No change
2	4.1.1.1 4.1.1.2 4.1.1.5	4.1.1.1 4.1.1.2 4.1.1.5	No change
3	4.1.1.3 4.1.1.4	4.1.1.3 4.1.1.4	No change
4		4.1.1.6	Lesson not required for Combined: Trilogy. Replace lesson with activities based around required practical activity 1, such as magnification calculations and use of the prefixes *centi*, *milli*, *micro* and *nano*.
5	4.1.2.1 4.1.2.2 4.1.2.3	4.1.2.1 4.1.2.2 4.1.2.3	No change
6	4.1.3.1 4.1.3.2 4.1.3.3	4.1.3.1 4.1.3.2 4.1.3.3	No change
7	4.2.1 4.2.2.1	4.2.1 4.2.2.1	No change
8	4.2.2.2 4.2.2.3	4.2.2.2 4.2.2.3	No change
9	4.2.2.4 4.2.2.5 4.2.2.6 4.2.2.7	4.2.2.4 4.2.2.5 4.2.2.6 4.2.2.7	No change
10	4.2.3.1 4.2.3.2	4.2.3.1 4.2.3.2	No change
11	4.3.1.1 4.3.1.2 4.3.1.3	4.3.1.1 4.3.1.2 4.3.1.3	No change
12	4.3.1.4 4.3.1.5	4.3.1.4 4.3.1.5	No change
13	4.3.1.6 4.3.1.7 4.3.1.8	4.3.1.6 4.3.1.7 4.3.1.8	No change
14	4.3.1.9	4.3.1.9 4.3.2.1 4.3.2.2	Monoclonal antibodies not required for Combined: Trilogy. Instead, provide an outline of how aspirin is extracted from willow. Then compare the discovery and development of digitalis (William Withering) and penicillin (Alexander Fleming) with modern drug development.
15		4.3.3.1 4.3.3.2	Omit lesson. Replace with activities that test the student's understanding of required practical activities 2 and 3 and graphical interpretation of data on non-communicable disease.
16	4.4.1.1 4.4.1.2 4.4.1.3	4.4.1.1 4.4.1.2 4.4.1.3	No change
17	4.4.2.1 4.4.2.2 4.4.2.3	4.4.2.1 4.4.2.2 4.4.2.3	No change
18	4.5.1 4.5.2	4.5.1 4.5.2.1 4.5.2.4	Control of body temperature not required for Combined: Trilogy. Replace with activities that test the student's understanding of required practical activity 6.
19		4.5.2.2 4.5.2.3	Omit lesson. Replace with activities that test the student's understanding of required practical activities 4 and 5.

BIOLOGY HIGHER

Mapping of lessons to AQA GCSE (9-1) Combined Science: Trilogy

Lesson	Specification reference		Suggested adaptation for Trilogy
	Combined	Biology	
20	4.5.3.1 4.5.3.2	4.5.3.1 4.5.3.2	No change
21		4.5.3.3	This section is not required for Combined: Trilogy. Omit lesson. Replace with review of the nervous and endocrine systems; focus on answering longer exam-style questions.
22	4.5.3.3 4.5.3.4	4.5.3.3 4.5.3.4	No change
23	4.5.3.5 4.5.3.6	4.5.3.5 4.5.3.6	No change
24		4.5.4.1 4.5.4.2	Omit lesson. Replace with activities on extracting and interpreting data from graphs showing hormone levels during the menstrual cycle.
25	4.6.1.1 4.6.1.2	4.6.1.1 4.6.1.2 4.6.1.3	Advantages and disadvantages of sexual and asexual reproduction not required for Combined: Trilogy. Replace this section of the lesson with an activity to interpret chromosome behaviour during meiosis and fertilisation.
26	4.6.1.3	4.6.1.4 4.6.1.5	DNA structure not required for Combined: Trilogy. Replace this part of the lesson with activities on the importance of understanding the human genome.
27	4.6.1.4 4.6.1.5 4.6.1.6	4.6.1.6 4.6.1.7 4.6.1.8	No change
28	4.6.2.1 4.6.2.3	4.6.2.1 4.6.2.3 4.6.3.3	Understanding of genetics not required for Combined: Trilogy. Replace this part of the lesson with further activities on inheritance using Punnett squares, family trees, and using direct proportion and simple ratios in genetic crosses.
29	4.6.2.4	4.6.2.4 4.6.2.5	Cloning not required for Combined: Trilogy. Replace this part of the lesson with activities to compare and contrast the use of selective breeding and genetic engineering in agriculture, including risks and benefits.
30	4.6.2.2	4.6.2.2 4.6.3.1 4.6.3.2	Theory of evolution and speciation not required for Combined: Trilogy. Replace with activities to help the student extract and interpret information from charts, graphs and tables such as evolutionary trees.
31	4.6.3.1 4.6.3.2 4.6.3.3 4.6.3.4	4.6.3.4 4.6.3.5 4.6.3.6 4.6.3.7	No change
32	4.6.4 4.7.1.1	4.6.4 4.7.1.1	No change
33	4.7.1.2 4.7.1.3 4.7.1.4	4.7.1.2 4.7.1.3 4.7.1.4	No change
34	4.7.2.1 4.7.2.2	4.7.2.1 4.7.2.2 4.7.2.3	Decomposition not required for Combined: Trilogy. Replace this part of the lesson with activities to interpret predator–prey cycles and interactions within food webs.
35	4.7.3.1 4.7.3.2	4.7.2.4 4.7.3.1 4.7.3.2	Impact of environmental change not required for Combined: Trilogy. Replace this part of the lesson with activities that test the student's understanding of required practical activity.
36	4.7.3.3 4.7.3.4 4.7.3.5 4.7.3.6	4.7.3.3 4.7.3.4 4.7.3.5 4.7.3.6	No change
37		4.7.4.1 4.7.4.2 4.7.4.3	Omit lesson. Replace with activities that test the student's ability to calculate mean, mode and median, and to plot and draw appropriate graphs in the context of required practical activity.
38		4.7.5.1 4.7.5.2 4.7.5.3 4.7.5.4	Omit lesson. Prepare some examples of synoptic questions for the students to practice bringing together learning from different areas of the specification.

Needs analysis

For parents and guardians

We have a tutor because…
(Briefly explain why you have employed a tutor.)

Where we are currently…
(Briefly explain the student's current progress. Do you have access to reports and predicted grades?)

For students

Use this space to tell your tutor about yourself.

I am…
Tell your tutor what type of person you think you are. Are you quiet or outgoing? Are you confident about your abilities?

I like…
Explain to your tutor how you like to work. Do you like to work independently or with more guidance? Do you like to write your answers down or talk through them first? Do you like to be creative?

How I feel about Biology…
Do you like Biology? Try to explain why or why not. What are your favourite and least favourite parts?

 **BIOLOGY** HIGHER

Needs analysis

Our goals

Work together to set small, achievable goals for the year ahead. Make them as positive as you can and don't limit your goals to areas of Biology – think about personal development too. Together, look back at this list often to see how you are progressing.

Tick off each goal when you've achieved it

In four weeks' time, I will…

☐ _____
☐ _____
☐ _____
☐ _____
☐ _____
☐ _____

In three months' time, I will…

☐ _____
☐ _____
☐ _____
☐ _____
☐ _____
☐ _____

By the time I sit my exam, I will…

☐ _____
☐ _____
☐ _____
☐ _____
☐ _____
☐ _____

Progress report

Fill in the boxes below with help from your tutor.

My strengths are...
Which areas of Biology do you think you've done well in recently? List at least three.

My favourite Biology topic is...
Which Biology topic is your favourite? It doesn't have to be the one you're best at!

because...

The areas of Biology I need to work on are...
In which areas of Biology do you think you need more practice?

To improve these areas, we are going to...
This space is for your tutor to explain how he/she is going to help you become confident in these areas.

End-of-lesson report

We have looked at last week's homework and my tutor thinks…
This space is for your tutor to give feedback on last week's homework.

Today, we worked on…
This space is for you to list all of the topics and skills that you and your tutor have worked on today.

I feel…
This space is for you to explain how you feel about today's lesson. Did you enjoy it? Do you feel confident?

My tutor thinks…
This space is for your tutor to explain how the lesson went.

At home this week, we can…
This space is for your tutor to explain what your homework is and to give you other ideas for extra revision and practice.

 **BIOLOGY HIGHER**

1 Diagnostic lesson

Learning objectives

- To explore how the student likes to learn and study
- To identify strengths and weaknesses in key skills
- To identify areas of the specification that may need more attention

Specification links

- Maths skills, investigative skills

Starter activity

- **All about you; 10 minutes; page 18**

  Explain that it's important to take an approach in tutor sessions that will suit the student and for this you need to know a bit about them. The first activity will explore how confident they are at speaking. Use prompt questions if necessary but allow ample thinking time, and use open prompts like 'tell me more about that'. The second activity explores skills and learning styles.

Main activities

- **Words and numbers; 15 minutes; page 19**

  Questions 1–5 explore essential maths skills. The purpose is to check understanding, not to teach skills, so move quickly through the tasks. The AQA GCSE (9–1) Biology webpages include excellent maths teaching materials that could be used in later sessions if there are gaps in understanding. The written question can be answered using Key Stage 3 or GCSE knowledge, depending on where the student is in the course. Look at the student's approach to written questions, use of key terminology and literacy skills.

- **Investigation skills; 15 minutes; page 20**

  Question 1 explores how confident the student is at planning investigations. Only bullet points are required. Allow sufficient thinking time. This is one of ten required practicals for AQA GCSE (9–1) Biology (seven for Combined Science). Allow the student to answer the units question independently. Question 3 looks at appropriate graphs. This should be discussed after the student has had time to spot mistakes and to annotate the graph.

Plenary activity

- **My favourite science lesson; 10 minutes**

  Ask the student to tell you about the best science lesson they have ever had. Ask open questions to find out why this lesson made an impact.

Homework activity

- **The story so far; 20 minutes; page 21**

  The specification topics should be read and ticks placed to show which parts of the course they have covered in class and what they are confident with. If they have not started it yet, use the first column to show learning at Key Stage 3.

Support ideas

- **Words and numbers** If the student struggles to begin the written question, ask them to write down any individual words that they know that are linked to the immune system or vaccines. Don't support them to reach a perfect answer – this isn't the point of the exercise.
- **Investigation skills** If the student struggles to suggest investigation methods, use prompt questions to help. How could we measure whether a balloon had more or less water in it over time? Ask them to draw the potato cylinders in a test tube to help visualise them.

Extension ideas

- **Words and numbers** If the student has strong maths skills, sketch a graph typical of rate of reaction over time, then ask them to explain how they would work out the initial rate of reaction (by drawing a tangent to the curve).
- **Investigation skills** Explore the student's knowledge of other required practical activities. For example, ask them to describe how they would measure the rate of photosynthesis in pondweed and what factors might affect this.

Progress and observations

BIOLOGY HIGHER

Starter activity: All about you

Time 10 mins

Learning objectives
- To explore the student's interests and how they like to learn and study

Equipment
none

Take a minute to talk to your tutor about yourself. You can talk about anything that's important to you – for example, your family and friends, your hobbies and interests, your favourite subjects at school, your heroes and hopes for the future.

It's important to try out new ways to learn, but, as a starting point, it is good to know how you rate your skills and preferences.

Score the statements by placing a tick (✓) in a box from 1 (definitely not me) to 5 (spot on – that's me!).

Don't think too hard, it's not a test – just go with your gut feelings!

	1 Not me	2	3	4	5 That's me!
I enjoy doing practical work in science.					
I am good at drawing and using graphs.					
I learn best by doing practical investigations or by building models.					
I learn best when the teacher explains how things work and we ask questions.					
I learn best from diagrams or from watching animations or videos.					
I am good at maths and always confident when doing calculations in science.					
I am good at writing and score well in longer exam-style questions.					
I am good at using long science words.					
I love to draw and use colour in my notes.					
I am well organised. I plan my homework carefully and it's always in on time.					
I work best under pressure.					

BIOLOGY HIGHER

Main activity: Words and numbers

Time 15 mins

Learning objectives
- To explore the student's maths skills
- To see how the student approaches a written question

Equipment
none

1. Explain to your tutor what these symbols mean: =, ≠, <, >, ≤, ≥

2. Standard form makes very big and very small numbers more manageable.

    a) Write these numbers in standard form.

      i) 12 760 000 _____   ii) 0.000876 _____

      iii) 563 million _____   iv) 0.05643 _____

    b) Convert these standard form numbers into decimal form:

      i) 6.45×10^4 _____   ii) 3.2×10^{-3} _____

3. What is 0.008809 written to three significant figures?

4. In science, data can be summarised in several ways. Calculate the mean, median and mode for this data.

Diameter of snails (mm)	21	27	20	22	26	21	24

Mean _____   Median _____   Mode _____

5. Rearrange the equation to show how to calculate magnification.

    Measured size = actual size × magnification

6. People can be vaccinated against some infectious diseases. Describe four ways in which vaccination prevents a person getting a disease.

BIOLOGY HIGHER

Main activity: Investigation skills   Time 15 mins

Learning objectives
- To explore the student's practical investigation skills

Equipment
none

1. Potato cells will take up or lose water depending on the strength of the solution they are placed in. A cork-borer can be used to take cylinders from a potato. You are given five different concentrations of sugar solution.

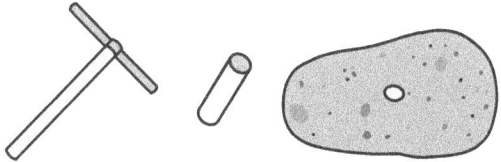

Make notes on how you could investigate the effect of sugar concentration on water uptake by potato cylinders.

What you would measure and how	Variables you would control

2. The metre is a unit of length. Name the following length units then convert 1 m into each unit.

_____ cm   _____ mm   _____ µm   _____ nm

3. The graph below was drawn to show the number of students with different eye colour in a class. It hasn't been drawn very well. Identify as many errors as you can – annotate the graph to show what is wrong.

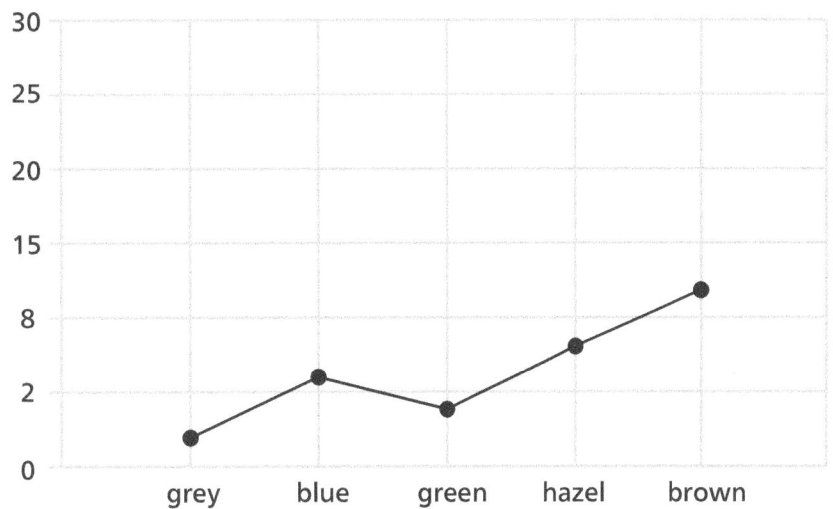

BIOLOGY HIGHER

Homework activity: The story so far

Time 20 mins

Learning objectives
- To explore the student's understanding of what you have covered so far

Equipment
none

Topic	Not covered yet	Confident	Not confident
eukaryotes and prokaryotes			
cell structure and specialisation			
culturing microorganisms			
mitosis and the cell cycle, stem cells			
diffusion, osmosis, active transport			
the human digestive system			
heart, blood vessels and blood			
coronary heart disease, health issues, lifestyle and cancer			
plant tissues, organs and systems			
infectious diseases (viral, bacterial, fungal, protist)			
human defence systems and vaccination			
antibiotics and painkillers, the discovery and development of drugs			
monoclonal antibodies			
plant diseases and defence			
photosynthesis and the uses of glucose			
aerobic and anaerobic respiration, metabolism			
homeostasis and negative feedback			
the human nervous system, the brain and eye			
control of body temperature			
human endocrine system			
control of blood glucose concentration			
water and nitrogen balance			
human reproduction, contraception and treating infertility			
plant hormones and their uses			
sexual and asexual reproduction and meiosis			
DNA, the genome and genetic inheritance (disorders, sex determination)			
the history of understanding genetics			
variation and selective breeding			
genetic engineering and cloning			
theory of evolution, speciation and extinction			
evidence for evolution, fossils and bacterial resistance			
classification of living organisms			
communities			
abiotic and biotic factors, and adaptations			
decomposition and how materials are cycled			
global warming, waste management, land use and deforestation			
trophic levels and pyramids of biomass			
food security and food production (farming, fishing and biotechnology)			

1 Answers

Starter activity: All about you

Student's own answers for discussion and information only. Use the answers to inform future learning, both by using preferred learning styles but also to develop and try out skills and approaches that are perhaps underused.

Main activity: Words and numbers

1. = equals; ≠ does not equal; < less than; > greater than; ≤ less than or equal to; ≥ greater than or equal to
2. a) i) 1.276×10^7
   ii) 5.63×10^8
   iii) 8.76×10^{-4}
   iv) 5.643×10^{-2}
   b) i) 64500
   ii) 0.0032
3. 0.00881
4. Mean 161/7 = 23; median = 22, mode = 21
5. Magnification = measured size ÷ actual size
6. Any four from: vaccine contains small quantities of dead or inactive (or antigens from) pathogen; stimulates white blood cells to produce antibodies; to produce antitoxins; antibodies kill (pathogen); (antibodies) stimulate phagocytosis; antitoxins neutralise poisons; memory cells remain; if the pathogen re-enters the body, correct antibodies produced quickly preventing infection

Main activity: Investigation skills

1. Measure length or mass at the start and end of the experiment. Use a ruler or mass balance. Calculate % change in length/mass. Control: starting length/mass of potato, volume of sugar solution, time left in solution, type/age of potato, temperature.
2. 100 cm (centimetres); 1000 mm (millimetres); 1 000 000 µm (micrometres); 1 000 000 000 nm (nanometres). It is better if standard form is used for longer numbers.
3. This is category data, so it should be a bar chart, not a line graph. The vertical axis has a non-linear scale. The maximum value on the vertical scale is not appropriate – the plotted data should take up at least half of the available space on the graph paper. The vertical scale could go from 0 to 15 in this case, for instance. There are no labels on the axes.

Homework activity: The story so far

Student's own answers for discussion and information only. Use this activity to find out which subject areas will need to be covered in more detail and those that may be moved through more quickly.

BIOLOGY HIGHER

2 Cell biology: Cell structure and microscopes

Learning objectives

- To describe the main differences in structure between eukaryotic and prokaryotic cells, and between animal and plant cells
- To explain how sub-cellular structures are related to their functions
- To understand how microscopes have developed over time
- To explain how electron microscopy has increased our understanding of sub-cellular structures
- To carry out calculations involving magnification

Specification links

- 4.1.1.1
- 4.1.1.2
- 4.1.1.5
- MS 1a, 1b, 2h

Starter activity

- **What's in a cell?; 5 minutes; page 24**

  Remind the student that there are differences between the cells of each major taxonomic group. Check that they understand that bacteria are prokaryotes, and that animals and plant cells are eukaryotic. Ask them to complete the table with a tick if a cell sub-structure is present and a cross if it is not. When complete, discuss any areas of uncertainty.

Main activities

- **Prokaryotic and eukaryotic cells; 15 minutes; page 25**

  Discuss the difference between prokaryotic and eukaryotic cells (see glossary). Ask the student to work through the activity sheet. Discuss the relative size of cells and cell sub-structures.

- **Subcellular structures and how to find them; 20 minutes; page 26**

  Ask the student to work through section 1 and explain things where needed. Then ask about their experience of using microscopes and what subcellular structures they could or couldn't see and why this might be. Explain the differences between magnification and resolution, perhaps using the idea of pixel size or print dots per inch to explain resolution. Ask the student to complete the activities, then discuss their answers.

- **Order of magnitude calculations and standard form; 5 minutes**

  A plant cell measures 50 µm in length. Ask the student to convert this into nanometres and millimetres and write the answers in standard form. Answer: 50 000 nm or 5×10^4 nm; 0.05 mm or 5×10^{-2} mm

Plenary activity

- **Sketch a bacterium; 5 minutes**

  Challenge the student to sketch and label a bacterial (prokaryotic) cell from memory and describe the function of different structures within it. For variety, this could be done large-scale on a window or on a chalk board.

Homework activity

- **Extended response – microscopes; 20 minutes; page 27**

  Ask the student to complete the six-mark exam-style question.

Support ideas

- **Prokaryotic and eukaryotic cells** Ask the student to describe how they would make a model of a cell. This will help those who struggle to visualise a cell in 3D.
- **Subcellular structures and how to find them** Ask the student to describe the steps needed to view a slide.

Extension ideas

- **Prokaryotic and eukaryotic cells** Fungi are another group, distinct from bacteria, plants and animals. Describe fungal cell features, such as DNA enclosed in nucleus, chitin cell wall, relatively large cells, and DNA in linear chromosomes. Ask the student to explain whether they are prokaryotic or eukaryotic.
- **Subcellular structures and how to find them** Sketch a basic cell. Ask the student to draw sections at different magnifications. Relate this to different planes of view and how cells can appear as different shapes in a cross section.

Progress and observations

BIOLOGY HIGHER

Starter activity: What's in a cell?

Time 5 mins

Learning objectives

- To review the main differences in structure between eukaryotic (animal and plant) and prokaryotic (bacteria) cells

Equipment

none

1. The table shows some different types of cell and a list of sub-structures they may contain. For each type of cell, fill in the table with a tick if a sub-structure is present and a cross if it is not.

Cell sub-structure	Eukaryotic cells		Prokaryotic cells
	Animal	Plant	Bacteria
nucleus			
cytoplasm			
large DNA loop			
cell wall			
cell membrane			
mitochondria			
ribosomes			
chloroplasts			
permanent vacuole filled with cell sap			
plasmids			

BIOLOGY HIGHER

Main activity: Prokaryotic and eukaryotic cells

Time 15 mins

Learning objectives

- To describe the main differences in structure between eukaryotic and prokaryotic cells, and between animal and plant cells
- To explain how sub-cellular structures are related to their functions

Equipment

none

1. Label the diagrams below as either 'eukaryotic cell (plant cell)' or 'prokaryotic cell (bacterium)'. Then label as many parts of each cell as you can.

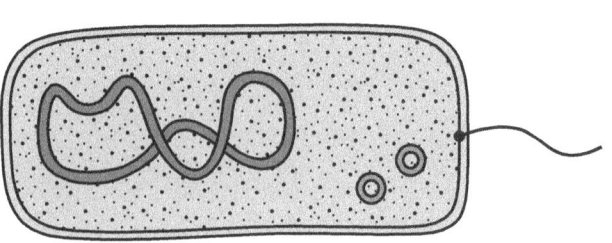
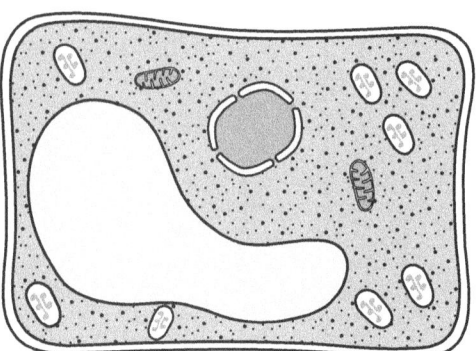

2. Why have ribosomes not been shown on these cell diagrams?
   _____

3. Fill in the gaps in the following sentences.

   Cells that have their genetic material enclosed in a nucleus are known as _____ cells. Plant and

   _____ cells are examples of this kind of cell. Prokaryotic cells such as bacterial cells are much

   _____ in size. The _____ material is not enclosed in a _____ . It is a

   single _____ loop and there may be one or more small rings of DNA called _____ .

   All prokaryotic and eukaryotic cells have both cytoplasm and a cell _____ . Bacteria have a cell wall

   made of murein and plant cells have a cell wall made of _____ . There is no cell wall surrounding

   _____ cells.

4. The actual length of the plant cell from which the drawing above was made was 50 µm. Draw a scale bar on the diagram to show a length of approximately 10 µm on the image.

5. The plant cell in the diagram measured 50 µm in length. Convert this into nanometres and millimetres and write your answers in standard form.
   _____
   _____

BIOLOGY HIGHER

Main activity: Subcellular structures and how to find them   Time 25 mins

Learning objectives

- To describe how the main sub-cellular structures are related to their functions
- To explain how electron microscopy has increased our understanding of subcellular structures
- To carry out calculations involving magnification

Equipment

- calculator

1. Draw one line from each part of a cell to a detail of its structure. Draw another line to link to its function.

nucleus	partially permeable	carries additional genes and may be passed on from one bacteria to another
cell membrane	large internal surface area for reactions	where aerobic respiration takes place, releases energy from food
mitochondria	contains chlorophyll pigment	controls what goes into and out of the cell
chloroplast	contains chromosomes made of DNA	traps the energy from sunlight
plasmid	small circular piece of DNA	stores the genetic material (in eukaryotes)

2. Light microscopes have been largely replaced by electron microscopes in research into cell structure. An electron microscope has much better resolving power than a light microscope. This has enabled biologists to find out about many more subcellular structures. Explain how the better resolution of electron microscopes allowed scientists to discover more subcellular structures.

_____

_____

_____

3. In the following calculations, use the formula: magnification = size of image ÷ size of real object.

   a) An image of a red blood cell in a book measures 2.1 cm in diameter. Red blood cells have a diameter of 7 µm. Calculate the magnification used for the image.

   _____

   b) You are given a drawing of an animal cell. The magnification is given as ×2000. The image measures 40 mm. What is the size of the cell? Give your answer in micrometres.

   _____

BIOLOGY HIGHER

Homework activity: Extended response – microscopes

Time 20 mins

Learning objectives

- To understand how microscopes have developed over time
- To explain how electron microscopy has increased our understanding of sub-cellular structures

Equipment

none

1. Read the following exam-style question.

   The table shows some information about microscopes.

	Light microscope	Electron microscope
Resolution	low resolution	very high resolution
Magnification	up to ×1500	more than ×500 000
Type of material	dead or live material can be viewed	dead material only
Ease of use	easy to use, cheap to run	training needed, expensive to run
Image colour	natural colour of cells can be seen	black and white image only

   Evaluate which type of microscope is the most useful for scientists who want to find out about cells.

   [6 marks]

   Use data from the table and your own knowledge to support your answer.

2 Answers

Starter activity: What's in a cell?

1. The student should have ticked as follows:
   Animal: nucleus, cytoplasm, cell membrane, mitochondria, ribosomes
   Plant: nucleus, cytoplasm, cell wall, cell membrane, mitochondria, ribosomes, chloroplasts, permanent vacuole
   Bacteria: cytoplasm, large DNA loop, cell wall, cell membrane, ribosomes, plasmids

Main activity: Prokaryotic and eukaryotic cells

1. The student should have included labels as follows: bacterial cell: plasmid, DNA loop, cell wall, cell membrane, cytoplasm, flagellum; plant cell: nucleus, cell wall, cell membrane, cytoplasm, mitochondria, large (permanent) vacuole, chloroplasts
2. Ribosomes are much smaller than other parts of the cell. It would not be possible to show any detail of them on a drawing of this scale (they could perhaps be drawn as 'dots').
3. Eukaryotic; animal; smaller; genetic; nucleus; DNA; plasmids; membrane; cellulose; animal
4. The scale bar should be approximately one fifth of the length of the plant cell.
5. 50 000 nm or 5×10^4 nm; 0.05 mm or 5×10^{-2} mm

Main activity: Subcellular structures and how to find them

1. Nucleus: contains chromosomes made of DNA, stores the genetic material (in eukaryotes)
   Cell membrane: partially permeable, controls what goes into and out of the cell
   Mitochondria: large internal surface area for reactions, where aerobic respiration takes place, releases energy from food
   Chloroplast: contains chlorophyll pigment, traps the energy from sunlight
   Plasmid: small circular piece of DNA, carries additional genes and may be passed on from one bacteria to another
2. Better resolution allows the image to be magnified more times and it will still stay clear/not become blurred. This means that it can be used to study cells in much finer detail.
3. a) 2.1 cm = 21 000 µm. Magnification is 21 000/7 = ×3000
   b) 20 µm (40 000/2000)

Homework activity: Extended response – microscopes

1. The following table provides guidance on what a Level 3, 2 or 1 answer to this question would look like and the number of marks each would attract.

L3	A detailed and coherent evaluation is provided that considers a range of relevant points, quotes relevant information from the table and comes to a conclusion consistent with the reasoning.	5–6 marks
L2	An attempt to describe relevant points which comes to a conclusion. The logic and use of information may be inconsistent at times but builds towards a coherent argument.	3–4 marks
L1	Discrete, relevant points made. The logic may be unclear and the conclusion, if present, may not be consistent with the reasoning.	1–2 marks
	Indicative content A conclusion is made as to which type of microscope is more useful in research about cells. Points that may be used in argument: • Resolution of EM is better due to shorter wavelength of electron beam than light rays/photons • Link made between resolution and better magnification/image not blurred at high magnification/so finer details can be seen • Magnification is many times higher for EM, so many cell sub-structures can only be seen with EM • Suggestion of structure that cannot be seen with LM e.g. ribosomes/detail of mitochondria • But EM may not be available to some scientists due to high costs/difficulty of use, idea that this may limit research • EM can't be used for living cells – so difficult to find out about cell processes/movement/changes over time • Only LM shows natural colours – suggests valid advantage, e.g. some processes might cause a colour change/this may help to distinguish cell structures	

3 Cell biology: Cell specialisation and differentiation

Learning objectives

- To explain how cells may be specialised to carry out a particular function
- To explain the importance of cell differentiation
- To compare differentiation in plants and animals
- To explain how the structure of different types of cell relates to their function

Specification links

- 4.1.1.3
- 4.1.1.4

Starter activity

- **Specialisation and differentiation; 5 minutes; page 30**
  Remind the student that there may be many types of cell within one organism, each performing different roles. Ask them to complete the activity. Discuss any misunderstanding of key terms. Ask about the functions of the cells listed by the student.

Main activities

- **Specialised plant and animal cells; 25 minutes; page 31**
  Work through the table with the student one cell at a time, asking them to complete it with concise notes. The muscle cell and phloem cell may be unfamiliar, but are on the specification. If time is short, discuss any familiar cells verbally and ask the student to complete this part of the table for homework instead of the mind-map.
- **Differentiation in plant and animal cells; 15 minutes; page 32**
  In question 1 ask the student to fill in the gaps to note the key learning points. In question 2 discuss the differences in differentiation over the lifetimes of plants and animals. Ask the student to give examples of how they know that some plant cells retain the ability to differentiate (see answers for examples). Perhaps also show pictures of tissue culture to produce whole plants from explants. Discuss the idea that some tissues have a layer of cells that divide then differentiate to produce new cells, for example in the skin.

Plenary activity

- **Learning review; 5 minutes**
  Ask the student to write down one thing they learned this session and one question they would like to ask. This could be done verbally, but allow 30 seconds of thinking time. Use this as a focus to review their learning.

Homework activity

- **Specialised cells summary; 40 mins; page 33**
  The student is asked to produce a mind-map to summarise the ideas covered in this lesson.

Support ideas

- **Specialised plant and animal cells** Ask the student to give verbal definitions of a cell, tissue, an organ and organ systems. Then ask them to give an example of each in animals.
- **Specialised plant and animal cells** Students sometimes struggle to understand the 3D shape of cells. Help the student by asking them to make models of various specialised cells out of modelling clay.

Extension ideas

- **Specialised plant and animal cells** Explain that the type of muscle cell described is found in voluntary muscle. Ask the student to suggest where in the body you might find involuntary muscle, where you don't have conscious control over contractions. Discuss the similar properties (contraction) of smooth, cardiac and skeletal muscle.
- **Differentiation in plant and animal cells** Plants continue to grow throughout their life in both length and girth. Ask the student to suggest how this happens. Show the student a picture from a textbook or the internet of the position of meristem tissue in plants. Ask students to describe similarities to and differences from growth in animals.

Progress and observations

BIOLOGY HIGHER

Starter activity: Specialisation and differentiation   Time 5 mins

Learning objectives
- To check your understanding of keywords for this topic

Equipment
none

1. Fill in the gaps to match the keywords with their definitions.

cell _____ = the job or purpose of a cell

cell _____ = how the cell is organised or put together; its shape and what is inside it

a _____ cell = a cell with features that allow it to do a particular job in an organism

cell _____ = the process by which a cell develops in order to do a specific job

> **Keywords:** differentiation; specialised; function; structure

2. Can you remember any examples of specialised plant or animal cells? Try to list three examples of each.

Specialised plant cells: _____
_____

Specialised animal cells: _____
_____

3. Put the following structures in order, starting with the largest and ending with the smallest.

organ   cell   system   tissue   organism

_____
_____

BIOLOGY HIGHER

Main activity: Specialised plant and animal cells   Time 25 mins

Learning objectives

- To explain how the structure of different types of plant and animal cells relates to their function within a tissue, organ, organ system or whole organism

Equipment

none

1. Complete table to describe the function and structure of some specialised cells. One has been done for you.

Specialised Cell	Function	Structure and explanation
red blood cell	carries oxygen around the body	1. contains haemoglobin, which binds to oxygen 2. bi-concave disc shape gives a large surface area for gas exchange 3. small and flexible; allows cell to fit through capillaries
muscle cell		
nerve cell		
root hair cell		
xylem		
phloem		

BIOLOGY HIGHER

Main activity: Differentiation in plant and animal cells

Time 15 mins

Learning objectives
- To explain the importance of cell differentiation
- To compare differentiation in plants and animals

Equipment
none

1. Fill in the gaps in these sentences.

   In the very early embryo, all _____ are unspecialised, so they have a similar structure. These cells _____ to produce more unspecialised cells. As an organism develops, cells begin to change or _____ to form different types of cells. As a cell differentiates, it gains different sub-cellular structures. These structures will help the cell to carry out a certain function – it has become a _____ cell.

2. Most types of animal cell differentiate at an early stage, whilst many types of plant cells keep the ability to differentiate throughout their lifespan. Think about what you know about plant and animal growth. Try to think of two examples that show that plant cells keep their ability to differentiate.

   _____
   _____
   _____

3. In mature animals, cell division is mainly restricted to repair and replacement of cells. Large-scale differentiation of cells, to produce whole new organs, for example, does not normally occur in animals. Try to think of two examples where cells in animals need to divide to produce replacement cells.

   _____
   _____
   _____
   _____
   _____
   _____
   _____
   _____

 **BIOLOGY HIGHER**

Homework activity: Specialised cells summary

Time **40** mins

Learning objectives
- To explain how cells may be specialised to carry out a particular function
- To explain how the structure of different types of cell relates to their function
- To explain the importance of cell differentiation
- To compare differentiation in plants and animals

Equipment
- pencil for sketching
- coloured pens or pencils

1. Sperm cells are specialised cells.

This is a description of a sperm cell and the adaptations that it has to perform its function.

> The head contains the genetic information and an enzyme to help penetrate the egg cell membrane. The middle section is packed with mitochondria for energy for movement. It has a tail that moves to propel the sperm to the egg.

a) What is the function of a sperm cell?

_____

_____

b) Draw a sperm cell and label the structures that allow it to carry out this function.

2. Produce a mind-map to summarise your learning about specialised cells and cell differentiation.

Use colour and diagrams – these will help you to remember what you have learned.

33

3 Answers

Starter activity: Specialisation and differentiation

1. Function; structure; specialised; differentiation
2. Student's own answers
3. Organism; system; organ; tissue; cell

Main activity: Specialised plant and animal cells

1. Muscle cell: Function: to produce movement
   Structure and explanation: contains special protein fibres; proteins contract and shorten the muscle fibre; contains lots of mitochondria which supply energy for contraction
   Nerve cell: Function: to pass electrical messages around the body
   Structure and explanation: very long cell so it can carry signals over longer distances; fatty sheath which insulates cell and speeds up impulse; has many small connecting projections called dendrites; this allows impulses to be passed to/from other cells
   Root hair cell: Function: to absorb water from the soil by osmosis and mineral ions by active transport
   Structure and explanation: long finger-like projection of cell increases surface area for absorption; thin projection of cell can penetrate between soil particles to reach water/minerals; cell contains lots of mitochondria which release energy for active transport of minerals
   Xylem: Function: transports water and mineral ions through the plant from roots to leaves (also provides support)
   Structure and explanation: long cell with narrow lumen so easier to carry water upwards over longer distances; no cytoplasm (dead cells) so water can move more freely through the cell; cell walls contain lignin which waterproofs the cell to reduce sideways loss of water (also provides strength to cell)
   Phloem: Function: transports sugars made in leaves to the rest of the plant (translocation)
   Structure and explanation: long narrow cells which are better to carry sugars over longer distances; no nucleus and fewer other organelles which allows sugar solution to move through the cytoplasm more freely; cytoplasm is continuous between adjacent cells (via sieve plate) which allows rapid movement of sugar solution from cell to cell; have companion cell with many mitochondria, which provides energy to transport sugars into phloem (into sieve tube element)

Main activity: Differentiation in plant and animal cells

1. Cells, divide, differentiate, specialised
2. Examples might include: continued growth of plants over their whole lifetime, whereas animals usually stop growing; growing whole plants from cuttings, pruning/hedge cutting and subsequent regrowth of leaves/stems
3. Examples might include: healing of cuts or wounds, and replacement of cells when they die, especially those with a short life-span such as skin cells, red blood cells or gut epithelium

Homework activity: Specialised cells summary

1. a) To carry the genetic material to the egg cell
   b) Students should draw a sperm cell with head, middle section and tail, and add labels. The labels should read:
      Head: enzyme sac and genetic material; middle: mitochondria; tail
2. Check that the student has included enough, but not too much detail to summarise the topic. Discuss the benefits of the organisation of knowledge, use of colour, etc, to help them to remember information.

 HIGHER

4 Cell biology: Culturing microorganisms

Learning objectives

- To describe how bacteria can be grown in the laboratory
- To explain why aseptic techniques must be used when culturing bacteria
- To describe how to investigate the effect of antibiotics or antiseptics on bacterial growth

Specification links

- 4.1.1.6
- MS 1b, 5c

Starter activity

- **Growing bacteria; 10 minutes; page 36**

  This starter activity reviews the basic principles of culturing bacteria. Use question 1 for initial verbal discussion. Then allow the student to work through questions 2–4. Use question 2 to gauge the student's practical knowledge. Expression in standard form (question 4) is required for higher level; spend time going through this if the student is unsure (see support ideas). Make sure the student understands the term 'mean division time'.

Main activities

- **Aseptic technique; 15 minutes; page 37**

  First, make sure that the student understands keywords such as 'uncontaminated'. Allow them two minutes thinking time, then fill in the table together. For the next part, ask them to think about where contamination might come from.

- **Investigations with antiseptics or antibiotics; 15 minutes; page 38**

  Note that investigating the effect of antiseptics or antibiotics on bacterial growth using agar plates and measuring zones of inhibition is a required practical for AQA GCSE (9–1) Biology students. This activity helps the student to prepare for an exam-style question, which they will complete for homework. For question 2, you may need to remind the student of the formula for how to calculate area of a circle (πr^2).

- **Patterns of growth; 5 minutes**

  Sketch graph axes of numbers against time. Ask the student to sketch the shape of the graph of bacterial population growth over time in a liquid culture. The names of growth phases, such as log, stationary and death phase, are not required for 4.1.1.6, but the general idea that rapid division will only occur while culture conditions are suitable is important.

Plenary activity

- **Traffic lights; 5 minutes**

  Go through the learning activities with the student and ask them to give a traffic light colour to each. Green = good understanding, amber = unsure and need to read over at home, red = still confused, need to revisit with tutor.

Homework activity

- **Exam question – effect of antiseptics; 15 minutes; page 39**

  The student is asked to write an answer to the exam-style question as planned during the tutor session.

Support ideas

- **Growing bacteria** If the calculation proves difficult, start by asking students to write down how many bacteria there are after every division period. Then give further examples of calculations using different starting populations and mean division times.
- **Patterns of growth** Further practice in converting into standard form may be required, for small and large numbers, for example: 0.0036 (3.6×10^{-3}), 152 000 (1.52×10^5).

Extension ideas

- **Growing bacteria** Ask students to suggest what types of nutrients should be present in a culture broth.
- **Investigations with antiseptics or antibiotics** Explain that a chemical must diffuse through the agar to reach bacteria and kill them. Ask what the effect of temperature, solubility and size of molecule might have on results.

Progress and observations

BIOLOGY HIGHER

Starter activity: Growing bacteria

Time 10 mins

Learning objectives
- To describe how bacteria can be grown in the laboratory
- To calculate the number of bacteria using the mean division time
- To present results using standard form

Equipment
- calculator

1. We talk about 'growing bacteria' in the laboratory. What do we actually mean by this?

   _____
   _____

2. Bacteria can be grown in a nutrient broth solution or as colonies on an agar gel plate. Add labels to the diagrams below to compare these two methods. Make your labels as clear and as specific as you can.

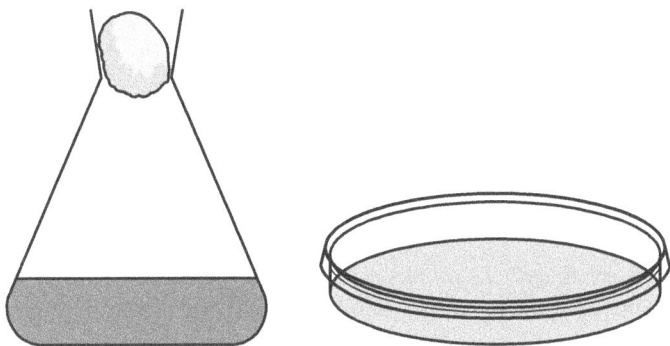

3. Bacteria multiply by simple cell division (binary fission). In the right conditions, this can be as often as once every 20 minutes. What sort of conditions would produce the fastest growth?

   _____
   _____

4. The starting population of bacteria in a liquid culture was 1.5×10^6. Calculate the number of bacteria that would be present in the population after three hours if the mean division time was 30 minutes. Write your answer in standard form.

   _____
   _____
   _____
   _____

36

BIOLOGY HIGHER

Main activity: Aseptic technique

Time 15 mins

Learning objectives
- To explain why aseptic techniques must be used when culturing bacteria

Equipment
none

1. A student wanted to prepare an uncontaminated culture of bacteria. The table below shows some steps that she carried out. Discuss the reason for each step with your tutor, then write down an explanation.

Step carried out	Reason
A. The glass Petri dishes and the culture medium (agar) were treated in a pressure cooker at 120 °C for 15 minutes.	
B. The lid of the Petri dish was secured with adhesive tape, but not completely sealed.	
C. The Petri dish was stored upside down.	
D. The inoculating loop used to transfer the chosen bacteria onto the media was passed through a Bunsen burner flame.	
E. The culture was incubated at 25 °C.	

2. Taking precautions to prevent contamination when culturing microorganisms is known as aseptic technique. How many other precautions can you think of, in addition to those in the table above, that would be good aseptic technique?

_____
_____
_____
_____
_____
_____
_____

37

BIOLOGY HIGHER

Main activity: Investigations with antiseptics or antibiotics

Time 15 mins

Learning objectives
- To describe how to investigate the effect of antibiotics or antiseptics on bacterial growth

Equipment
- pencil

1. Read the following exam-style question.

> A student wants to compare the effect of three different brands of antiseptic solution on bacterial growth.
> The brands were labelled A, B and C.
> The student was supplied with agar plates on which bacteria were already growing. Outline a plan the student could use to investigate the relative effect of the three antiseptics. Your plan should include the apparatus used and consider safety.
>
> [6 marks]

In the space below, make a mind map to show what you know about this practical.

Think about the independent variable, dependent variable, controlled variables, any control that should be set up, equipment and safety. Include a diagram to show what the set-up might look like.

2. A student carries out this experiment and measures the average diameter of the clear zone of one of the antiseptic discs as 12.4 mm. Work out the area of the zone of inhibition.

___
___
___
___
___

BIOLOGY HIGHER

Homework activity: Exam question – effect of antiseptics

Time 15 mins

Learning objectives
- To describe how to investigate the effect of antibiotics or antiseptics on bacterial growth

Equipment
none

1. Write an answer for the following exam-style question.

> A student wants to compare the effect of three different brands of antiseptic solution on bacterial growth. The brands were labelled A, B and C. The student was supplied with agar plates on which bacteria were already growing. Outline a plan the student could use to investigate the relative effect of the three antiseptics. Your plan should include the apparatus used and consider safety.
>
> **[6 marks]**

4 Answers

Starter activity: Growing bacteria

1. 'Growing bacteria' refers to bacteria dividing and the population of bacteria increasing in number, not each bacterium increasing in size. Culturing is a better term.
2. Labels could include: liquid nutrient broth, solid agar gel, bacteria grown suspended in broth, bacteria grown on gel surface, Petri dish, conical flask and cotton wool versus petri dish lid to prevent contamination
3. For rapid growth, bacteria require enough nutrients (carbohydrate, nitrogen/protein, vitamins/minerals) and a suitable temperature (depends on species; they need a warm temperature, but not so hot to kill them).
4. 9.6×10^7

Main activity: Aseptic technique

1. A: To kill any unwanted bacteria that might be present; if these were cultured they could be harmful to the scientist, or might affect the results of the experiment
   B: So that the dish cannot be opened, which prevents contamination by microorganisms in the air; it is not sealed so that oxygen can still enter
   C: Stops condensation that forms on the lid from dripping onto the cultures and disturbing the colonies
   D: Heat from the flame kills any possible contaminating bacteria on the loop and stops them being transferred to the culture
   E: Warm enough to speed up the division time, but not close to human body temperature so less likely that harmful bacteria that cause disease in humans will survive
2. Suggestions might include: wash hands before and after, disinfect bench, use a Bunsen burner flame to create upwards air current, only open Petri dish lid a small amount, heat sterilise/autoclave cultures and equipment after use

Main activity: Investigations with antiseptics or antibiotics

1. The student should be encouraged to provide as much of their own detail as possible.
   Independent variable: antiseptics A, B, C on separate paper discs
   Dependent variable: measure diameter or area of clear zone
   Controlled variables: same incubation temperature, amount of antiseptic, size of paper disc
   Control: paper disk with distilled water in place of antiseptic
   The mark scheme for the homework will be a useful prompt, but should not be shared with the student at this stage.
   Note that a set volume of antiseptic or water could be added to a well cut into in the agar as an alternative to paper discs.

   The following table provides guidance on what a Level 3, 2 or 1 answer to this question would look like and the number of marks each would attract.

L3	A coherent method is described with relevant detail, which demonstrates a good understanding of the relevant scientific techniques and procedures. The steps in the method are logically ordered with the dependent and control variables correctly identified. The method would lead to the production of valid results.	5–6 marks
L2	The bulk of a method is described with mostly relevant detail, which demonstrates a reasonable understanding of the relevant scientific techniques and procedures. The method may not be in a completely logical sequence and may be missing some detail.	3–4 marks
L1	Simple statements are made which demonstrate some understanding of some of the relevant scientific techniques and procedures. The response may lack a logical structure and would not lead to the production of valid results.	1–2 marks

2. Use πr^2. Area = $120.8\,mm^2$. The clear area is a zone of inhibition – bacteria have been killed or cell division prevented.

Homework activity: Exam question – effect of antiseptics

Student's own work. Examples include:
Named apparatus: filter paper discs (or cork borer); ruler or graph paper (to measure); forceps or tweezers; (dropper) pipette/balance;
Safety: aseptic technique; reference to lid: keep over agar/tape closed; sterilisation of equipment or used cultures;
Method: same volume of each antiseptic put on paper disc; place discs on top of bacterial culture; repeats for each antiseptic, A, B and C; control: a disc with distilled water only; incubate all at 25°C; for a set time (such as 24 hours); measure the average diameter or area of clear zones; bigger clear zone will indicate more effective antiseptic
See mark scheme for level guidance.

5 Cell biology: Cell division

Learning objectives

- To understand that chromosomes are made of DNA and carry genes
- To describe the events of the cell cycle and mitosis
- To understand the function of mitosis
- To describe the function of stem cells in animals and plants
- To describe the process of therapeutic cloning
- To evaluate the risks and benefits of stem cell therapies

Specification links

- 4.1.2.1
- 4.1.2.2
- 4.1.2.3

Starter activity

- **Chromosomes and cell division; 5 minutes; page 42**

  In the first question, the statement should be no more than two sentences. The true/false activity provides an opportunity to check the student's knowledge and understanding across the topics of this lesson. Ask the student to convert false statements into true ones, either verbally or in a written sentence.

Main activities

- **The cell cycle and mitosis; 15 minutes; page 43**

  Students do not need to know the different phases of mitosis. Ask students to complete questions 1–4 and discuss. Emphasise the importance of mitotic cell division for multicellularity.

- **Stem cells; 15 minutes; page 44**

  Ask the students to complete question 1, then discuss. Continue, one question at a time. Remind them that cells from human embryos can be cloned and made to differentiate into most different types of human cells. For question 4 discuss what types of specialised cell would be needed (pancreatic glucose producing cells/beta cells, or nerve cells) and how this might be achieved. Discuss the process of therapeutic cloning in preparation for the homework.

- **Risks and benefits of using stem cells in medical treatments and research; 10 minutes**

  The student should be able to evaluate the practical risks and benefits, as well as social and ethical issues, of the medical use of stem cells. Ask the student to suggest one practical benefit, one ethical issue and one social issue associated with using stem cells. A practical benefit might be in treating cancer patients with stem cells to replace cells killed during treatment. An ethical issue might be whether it is right to use spare embryos from IVF to reduce suffering when they would be discarded anyway. A social issue might be that many people disagree with the use of embryonic stem cells on religious grounds, or view destroying an embryo as murder.

Plenary activity

- **Guess the word; 5 minutes**

  Read out some of the definitions from the glossary on page 245. The student must suggest which words are being described.

Homework activity

- **Therapeutic cloning; 15 minutes; page 45**

  The student needs to complete the work on the sheet. The diagram in question 1 should include sketches of cells.

Support ideas

- **Chromosomes and cell division** Sketch a diagram to show the relationship if there is any confusion.
- **Stem cells** Show pictures of longitudinal sections through root tips and transverse section through stems and point out the position of meristem tissue. Ask the student to suggest the types of tissue that might be made.

Extension idea

- **Stem cells** Discuss different levels of potency: totipotent, pluripotent and multipotent. Ask for suggested examples.

Progress and observations

BIOLOGY HIGHER

Starter activity: Chromosomes and cell division   Time 5 mins

Learning objectives
- To understand the relationship between chromosomes, DNA and genes
- To recall basic facts about the cell cycle and stem cells

Equipment
none

1. Write a statement about the structure of generetic material that connects the following words:

    | DNA | cell | genes | chromosomes | nucleus |

    _____
    _____
    _____

2. Read the following statements. Circle T if a statement is true or F if it is false.

   a) In body cells the chromosomes are normally found in pairs.   T F

   b) Most human cells contain 23 chromosomes.   T F

   c) After mitosis, the cell has double the amount of genetic material. T F

   d) Mitosis is a type of cell division.   T F

   e) A stem cell is a cell which is specialised to produce one type of cell. T F

   f) Stem cells in plants can differentiate into any type of plant cell, throughout the life of the plant. T F

   g) Stem cells in plants are found in the meristem tissue.   T F

   h) In therapeutic cloning, an embryo is produced with the same genes as the patient. T F

BIOLOGY HIGHER

Main activity: The cell cycle and mitosis   Time 15 mins

Learning objectives
- To describe the stages of the cell cycle, including mitosis
- To understand the function of mitosis

Equipment
- pencil

1. Cells divide in a series of stages called the cell cycle. During the cell cycle the genetic material is doubled and then divided into two identical cells. Read the statements below. Decide what order they should go in, then add them as labels in the appropriate place on the diagram.

A. The cell grows. The number of subcellular structures, such as ribosomes and mitochondria, increases.	D. Further growth occurs. DNA is checked for errors.
B. One set of chromosomes is pulled to each end of the cell, forming into two nuclei.	E. Two identical cells form.
C. DNA replicates (genetic material is copied).	F. The cytoplasm and cell membranes divide.

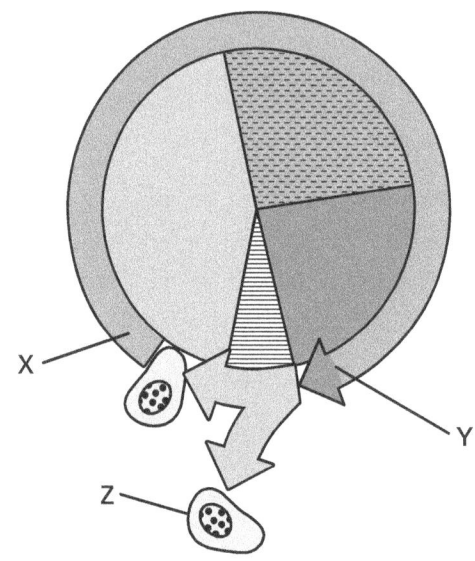

2. Add a label to this diagram to show where mitosis is occurring.

3. Draw what a chromosome would look like at point X and at point Y in the cell cycle.

4. What may happen to the new cells present at point Z?

BIOLOGY HIGHER

Main activity: Stem cells

Time 15 mins

Learning objectives
- To describe the function of stem cells in embryos, in adult animals and in the meristems in plants

Equipment
- pencil

1. A stem cell is an unspecialised cell that can divide to give many more cells which may then differentiate into specialised cells. Write down the function of each of the types of stem cells in the table.

Stem cell type	Function
stem cells in embryos	
stem cells in adult animals	
stem cells in the meristems of plants	

2. What is special about the cells within the early embryo?

_____

_____

3. One type of stem cell in adult bone marrow can divide and differentiate to produce blood cells, including red blood cells, white blood cells and platelets. Sketch a diagram to explain this.

4. Name two human conditions that can be treated using stem cells.

_____

5. In therapeutic cloning, an embryo is produced with the same genes as the patient. Stem cells from the embryo are then used to treat the patient. What is the advantage of using therapeutic cloning compared with other types of stem cell therapy?

_____

_____

_____

6. Stem cells from meristems in plants can be used to produce clones of plants quickly and cheaply. Give two examples of how cloning plants from meristems can be useful.

_____

_____

_____

_____

BIOLOGY HIGHER

Homework activity: Therapeutic cloning

Time 15 mins

Learning objectives
- To describe the process of therapeutic cloning
- To evaluate the risks and benefits of stem cell therapies

Equipment
- pencil

1. **Read the following description of therapeutic cloning.**

> In therapeutic cloning, the nucleus of a body cell from the patient is removed. This nucleus is then transferred to an egg cell that has had its own nucleus removed. The egg cell with the patient's nucleus is stimulated to divide to produce an embryo. Stem cells are removed from this embryo and used to treat the patient. The rest of the embryo is disposed of.

Draw a diagram to show the steps in therapeutic cloning. Use all the information in the description above to help you.

2. **Therapeutic cloning has not yet been used in treatments. Some people have concerns about this potential therapy. Give one ethical argument for and one ethical argument against the use of therapeutic cloning.**

For: _____

_____

_____

Against: _____

_____

_____

5 Answers

Starter activity: Chromosomes and cell division

1. For example: The nucleus of a cell contains chromosomes made of DNA molecules; each chromosome carries many genes.
2. a) T
   b) F: Most human cells contain 46 (23 pairs of) chromosomes.
   c) F: Mitosis divides the genetic material into two daughter cells.
   d) T
   e) F: A stem cell is an undifferentiated cell which divides to produce similar cells which may then differentiate into other cells.
   f) T
   g) T
   h) T

Main activity: The cell cycle and mitosis

1. A, C, D, B, F, E. Labels for A, C, D should be spaced out along the cell cycle line between the end and beginning of mitosis; B, F, E should be in order through mitosis
2. Mitosis labelled on dividing arrow
3. At point X the chromosome should be represented by a single linear structure, at Y it should be represented by a double structure (an elongated X shape) as the DNA has replicated.
4. The cell may begin the cell cycle and divide again (possibly after a resting stage), or it may differentiate and not divide again.

Main activity: Stem cells

1. In embryos, stem cells differentiate to give all the different tissue types that form the whole baby.
   In an adult animal, stem cells replace cells lost through damage or disease, and can differentiate to give only a limited number of different cell types within the same tissue.
   Meristem cells can differentiate to produce any type of plant cell for repair, to produce new organs (leaves, flowers, and so on) or for continued growth throughout the entire life of the plant.
2. Cells from the early embryo can differentiate into any type of cell, and even into an entire new individual if removed. This ability reduces as the embryo develops.
3. Diagram might show a stem cell dividing by mitosis, to produce several cells, each differentiating into a different blood cell
4. Diabetes and paralysis
5. The stem cells would not be rejected because they have the same genes as the patient.
6. Rare plant species can be cloned to protect them from extinction. Crop plants with special features, such as disease resistance, can be cloned to produce large numbers of identical plants for farmers.

Homework activity: Therapeutic cloning

1. The student's diagram should cover all the steps described in the text. It should include cell diagrams to show nuclear transfer and show one embryo cell dividing to produce several (a ball of cells). Differentiation of these cells should be labelled.
2. Any suitable answer that considers the rights and wrongs of the process; for example, for: the treatment may save the patient's life or reduce pain, while the embryo is unlikely to experience pain; against: some people might think that the embryo is a potential human being, and as it will be destroyed some people may consider this to be murder.

6 Cell biology: Transport in cells

Learning objectives

- To understand the processes of osmosis, diffusion and active transport
- To explain the factors that affect diffusion rate
- To calculate and compare surface area to volume ratios
- To explain how some surfaces are adapted for exchanging materials
- To explain the effect of solute concentration on the mass of plant tissue
- To explain the need for transport systems and specialised exchange surfaces

Specification links

- 4.1.3.1
- 4.1.3.2
- 4.1.3.3
- MS1c, 4c, 5c

Starter activity

- **Moving into and out of cells; 5 minutes; page 48**

  Allow the student to complete both questions on the sheet. Use question 1 to check understanding of the key terms. Emphasise the random movement of particles in all directions in diffusion. Discuss the answers for question 2 and ask the student to consider if there would be any differences in plant cells. Check their understanding that, in mammals, the waste product urea diffuses from cells into the blood plasma for excretion in the kidneys.

Main activities

- **Diffusion rate and exchange surfaces; 15 minutes; page 49**

  The questions can all be done verbally, except for completing the table.

- **Osmosis and active transport; 25 minutes; page 50**

  Students may understand the osmosis model diagram seen in textbooks, but may struggle to reproduce it. Question 2 shows results from a required practical. These will be used for the homework, so check that the student is familiar with the method. The answers to question 3 cover the required examples from the specification.

- **Comparing cell transport processes; 10 minutes**

  Discuss the differences (and similarities) between diffusion, osmosis and active transport. Consider energy requirement, use of carrier molecules, direction compared to concentration gradient, types of substances that might be transported.

Plenary activity

- **Revisit cell diagram; 5 minutes**

  Ask the student to annotate the diagram from the starter activity with information that they have learned in the session.

Homework activity

- **Osmosis investigation; 20 minutes; page 51**

  The student needs to plot a graph using the results table from the main activity. Salt concentration should be on the *x*-axis and % change in mass on the *y*-axis. Point out that there are negative and positive values for % change in mass, so the *x*-axis will be drawn at the 0% point and not at the lower margin.

Support ideas

- **Diffusion rate and exchange surfaces** Spray scent or air freshener on the other side of the room, after checking there's no asthma risk. Discuss how the scent particles reach the student's nose. Key terms – kinetic energy, random, spread out.
- **Osmosis and active transport** Draw diagrams to model diffusion, osmosis and active transport across a cell membrane.

Extension ideas

- **Diffusion rate and exchange surfaces** Ask the student to sketch a graph of the relationship between temperature and diffusion rate.
- **Osmosis and active transport** Ask the student to sketch a diagram to show how carrier proteins in the cell membrane are involved in active transport

Progress and observations

Starter activity: Moving into and out of cells

Time 5 mins

Learning objectives
- To understand how substances can move into and out of cells

Equipment
- pencil

1. **Substances may move into and out of cells across the cell membranes via diffusion, active transport or osmosis. Fill in the gaps below to remind yourself of the definitions of these processes (a word bank is provided).**

   Diffusion is the spreading out of the particles of any substance in solution, or particles of a gas, resulting in a net movement from an area of _____ concentration to an area of _____ concentration. Water may move across cell membranes via osmosis. Osmosis is the diffusion of water from a dilute solution to a _____ solution through a partially _____ membrane.

   Active transport moves substances from a more _____ solution to a more _____ solution (against a concentration gradient). This requires energy from _____ .

   > Word bank (words can be used more than once): concentrated, higher, permeable, dilute, respiration, lower

2. **The diagram below represents an animal cell. Draw labelled arrows to show what substances must be moved into animal cells and what must be removed from them. Discuss with your tutor how these substances might travel across the cell membrane.**

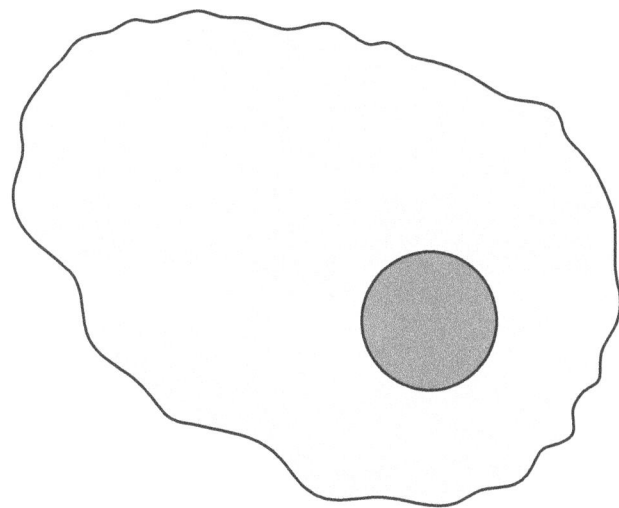

 **BIOLOGY HIGHER**

Main activity: Diffusion rate and exchange surfaces   Time 15 mins

Learning objectives

- To explain the factors that affect diffusion rate
- To calculate and compare surface area to volume ratios
- To explain the need for transport systems and exchange surfaces
- To explain how some surfaces are adapted for exchanging materials

Equipment

- pictures (textbook or web-based) of small intestine cells, fish gills, roots and leaves

1. There are three key factors that affect the rate of diffusion. These are:

   - the difference in concentrations (concentration gradient)
   - the temperature
   - the surface area of the membrane

   For each factor, explain to your tutor how it affects the rate of diffusion. Think of any keywords that are important.

2. A student carried out an experiment to compare diffusion rates in large and small animals. The student used cubes of agar jelly as model animals. The jelly contained an indicator which changes from green to red with acid. The jelly only changes colour where acid has diffused into the block.

    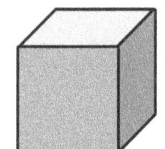

   small model animal   large model animal

   a) Complete the table.

Size	Cube dimensions (mm)	Surface area (mm²)	Volume (mm³)	Surface area to volume ratio
small	5 × 5 × 5	150		1.2 : 1
large	10 × 10 × 10		1000	

   b) After two minutes in acid, the student removed the cubes and cut them open. The small cube was red all the way through. The large cube was red only in the outer 3 mm. Use results from the table to explain these findings.

   _____

3. Most multicellular organisms need specialised exchange surfaces such as gills or digestive systems, while single-celled organisms do not. Explain this difference.

   _____

4. What other feature must multicellular organisms have to improve supply to and from body cells?

   _____

   The effectiveness of an exchange surface can be increased by having these features:

   - efficient blood supply
   - a membrane that is thin, to provide a short diffusion path
   - ventilation of gas exchange surfaces
   - a large surface area

5. Explain to your tutor how the following are adapted for exchanging materials: small intestine (mammals), lungs (mammals), gills (fish), roots, and leaves. Your tutor may show you pictures.

Main activity: Osmosis and active transport

Time 25 mins

Learning objectives
- To understand the processes of osmosis and active transport
- To explain the effect of solute concentration on the mass of plant tissue

Equipment
- pencil
- calculator

1. Sketch a diagram to demonstrate how water moves by osmosis across a partially permeable membrane.

   Use arrows to show particle movement. Your diagram should include these labels: solute molecule; water molecule; partially permeable membrane.

2. A student carried out an investigation into the effect of different concentrations of salt solution on the mass of pieces of potato. The results are given below. Complete the missing values and work out the percentage increase or decrease in mass.

Salt concentration (mol dm^{-3})	Initial mass (g)	Final mass (g)	Difference in mass (g)	Percentage change in mass (%)
0.0	2.63	2.88	0.25	9.5
0.2	2.57	2.72		
0.4	2.75	2.77	0.02	0.7
0.6	2.47	2.38	−0.09	−3.6
0.8	2.60	2.42		
1.0	2.59	2.40	−0.19	−7.3

3. In active transport, special carrier molecules in the membrane move substances against a concentration gradient. This requires energy from respiration. Give two examples where active transport must be used, one in plants and one in animals.

Plants: _____

_____

_____

Animals: _____

_____

_____

BIOLOGY HIGHER

Homework activity: Osmosis investigation

Time 20 mins

Learning objectives
- To draw and interpret graphs
- To explain the effect of solute concentration on the mass of plant tissue

Equipment
none

1. A student carried out an investigation into the effect of different concentrations of salt solution on the mass of pieces of potato. Potato pieces were weighed, placed in the solutions for 24 hours and then re-weighed. Results are shown in the table.

Salt concentration (mol dm^{-3})	Percentage change in mass (%)
0.0	9.5
0.2	5.8
0.4	0.7
0.6	–3.6
0.8	–6.9
1.0	–7.3

Plot a suitable graph to display these results.

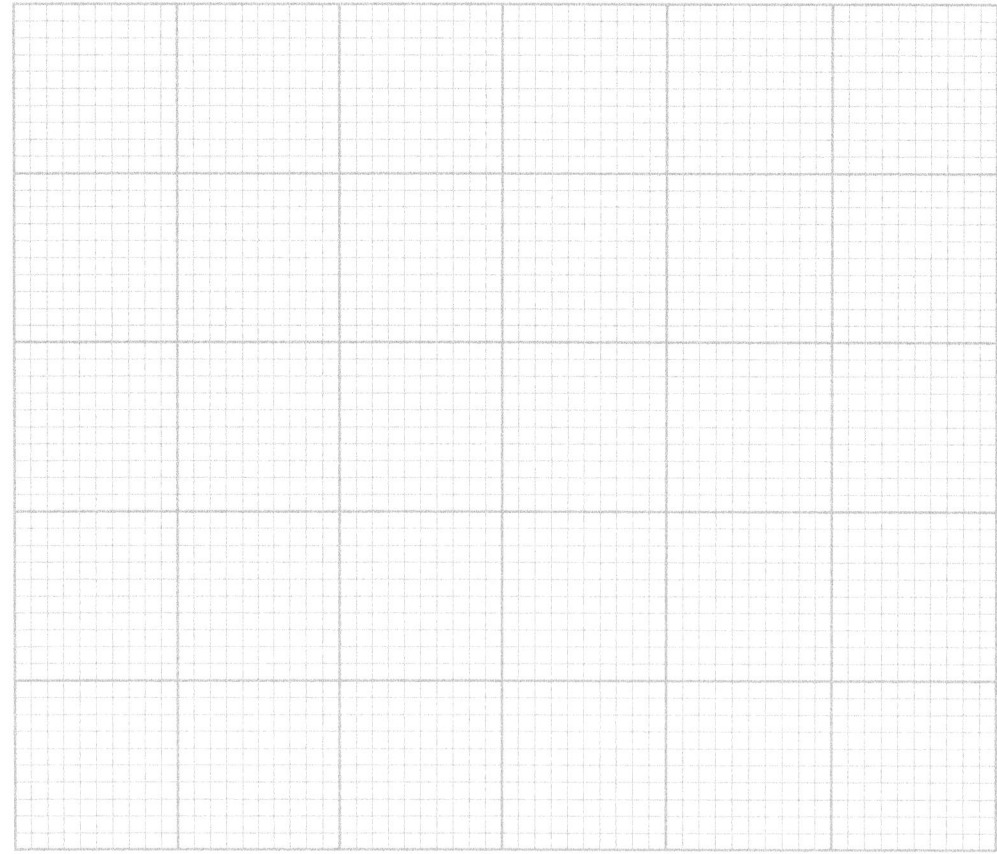

2. Use your graph to estimate the concentration of the solution inside the potato cells.

51

6 Answers

Starter activity: Moving into and out of cells

1. Higher, lower, concentrated, permeable, dilute, concentrated, respiration
2. Arrows in: oxygen, food/sugar/amino acids, ions, water; arrows out: CO_2, waste/urea, water

Main activity: Diffusion rate and exchange surfaces

1. Increasing the concentration gradient increases the rate of diffusion. Particles spread out from a point by random motion, so there are more particles available to move out from an area of high concentration than there are in the opposite direction. Higher temperatures provide greater (kinetic) energy to particles which move faster, increasing the rate of diffusion. The greater the surface area, the greater the rate of diffusion, as there will be a greater surface across which particles can pass.
2. a) 125; 600; 0.6 : 1
   b) The smaller cube has a larger surface area to volume ratio, so the acid can diffuse into the centre of the cube more quickly. Emphasise the similarities with living organisms.
3. Single-celled organisms are very small and so have a relatively large surface area to volume ratio. This allows sufficient transport of molecules into and out of the cell by diffusion to meet the needs of the organism. Multicellular organisms are larger and have a smaller surface area to volume ratio. Specialised exchange surfaces increase this surface area so that diffusion of gases or food is increased.
4. A transport system, such as the circulatory system in animals or a vascular system (xylem and phloem) in plants
5. The student should describe the adaptations provided where relevant, relating these to pictures they are shown.

Main activity: Osmosis and active transport

1. The student's diagram should show understanding that water molecules can move in both directions, that net movement is from dilute to concentrated solution, and that solute molecules cannot pass through the membrane.
2. 0.15, 5.8%; −0.18, −6.9%
3. Plants: active transport allows mineral ions to be absorbed into plant root hairs from very dilute solutions in the soil; plants require ions for healthy growth
   Animals: sugar molecules are absorbed from lower concentrations in the gut into the blood which has a higher sugar concentration; sugar molecules are used for cell respiration

Homework activity: Osmosis investigation

1. Correct scales chosen for axes (must maximise the area of the grid used)
   Both axes correctly labelled
   All points correctly plotted
   Smooth line joining the points or points joined with ruler as a suitable approach in this case
2. 0.44 mol dm^{-3}

BIOLOGY HIGHER

7 Organisation: The digestive system and enzymes

Learning objectives

- To understand relationships between cells, tissues, organs and systems
- To understand how the digestive system functions
- To explain enzyme structure and action, and the effects of pH and temperature on the rate of reaction
- To describe the function of a range of digestive enzymes and bile
- To describe how qualitative reagents are used to test for food substances

Specification links

- 4.2.1
- 4.2.2.1

Starter activity

- **Organisation and the digestive system; 5 minutes; page 54**

  Ask the student to complete the tasks independently, then discuss their answers. Check that they have retained knowledge of the digestive system studied in Key Stage 3 science on the outline functions of the main organs and the purpose of digestion. Mention briefly the position of the liver.

Main activities

- **Enzymes and digestion; 15 minutes; page 55**

  The three questions should be completed fairly quickly and should not be challenging to higher students. Question 1 checks understanding of the lock and key theory. Question 2 covers the production and function of the enzymes listed on the specification. In question 3 make sure that students understand why it is important that bile increases the surface area of fat droplets and the link between bile and lipase action.

- **Rates of enzyme reactions; 15 minutes; page 56**

  Explain that rates of reactions can be measured using the amount of substrate used, or product made, divided by time. Complete the activities one at a time and check understanding before continuing.

- **Tests for food substances; 10 minutes**

  Discuss food tests with the student by asking a series of questions. Ask the student to record information using a spider diagram as you proceed. Include: Benedict's test for sugars; iodine test for starch; and Biuret reagent for protein. Students need to know the name of the reagent, a very simple outline of the test (such as heat for Benedict's) and the colour change for a positive result. This is a required practical.

Plenary activity

- **Highlights; 5 minutes**

  Ask the student to look over the work that has been completed in the session and use a highlighter to mark the five most important things that they must remember or revisit.

Homework activity

- **Amylase investigation; 15 minutes; page 57**

  This investigation is a required practical on the specification. Check that the student is familiar with using a spotting tile.

Support ideas

- **Organisation and the digestive system** Discuss the diagram on the starter sheet and add labels on organ function.
- **Enzymes and digestion** Sketch a diagram of the action of bile and lipase on fat droplets to support understanding.

Extension ideas

- **Enzymes and digestion** Explain, with diagrams, how salivary amylase converts starch to maltose (a disaccharide) and then amylases in the gut break this down further into glucose.
- **Rates of enzyme reactions** Ask the student to explain how rate of reaction can be calculated from the gradient of the graphs.

Progress and observations

BIOLOGY HIGHER

Starter activity: Organisation and the digestive system

Time 5 mins

Learning objectives
- To understand the relationships between cells, tissues, organs and organ systems
- To describe the digestive system as an example of an organ system in which several organs work together to digest and absorb food

Equipment
none

1. Levels of organisation are described below. Draw lines to match up the boxes on the left with the correct definition.

cell	A. a group of cells with a similar structure and function
tissue	B. the basic building block of all living organisms
organ	C. group of organs which work together to carry out a specific function
organ system	D. an entire living thing, made up of all the other parts
organism	E. aggregation of tissues performing specific functions

2. List the organs, in order, that food would pass through after entering the mouth. Using the diagram below to help you. When you have finished discuss with your tutor what happens to the food at each stage.

54

BIOLOGY HIGHER

Main activity: Enzymes and digestion Time 15 mins

Learning objectives
- To explain enzyme structure and action
- To describe the production and action of bile and a range of digestive enzymes

Equipment
- pencil

1. Enzymes are biological catalysts which speed up chemical digestion. The diagram below is a model of the lock and key theory of how enzymes work.

   Add the labels: 'substrate', 'enzyme' and 'active site' to this diagram. Complete the diagram to show how the enzyme digests (breaks down) its substrate. Add the labels: 'enzyme–substrate complex' and 'products' to the diagram.

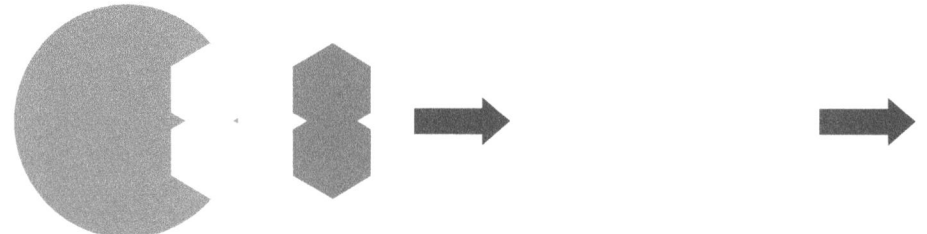

2. Digestive enzymes convert food into small soluble molecules that can be absorbed into the bloodstream. Some examples of these enzymes and their action are shown in the table. Complete the blank boxes.

Enzyme	Site of production	Digestion reaction that is catalysed Food → product (small soluble molecule)	How the products are used in the body
carbohydrase	salivary glands, pancreas, small intestine	carbohydrates → simple sugars	to build new carbohydrates
amylase (a type of carbohydrase)	salivary glands, pancreas, small intestine	→	
protease		proteins → amino acids	
lipase	pancreas, small intestine	→	to build new lipids (fats and oils)

3. Bile is not an enzyme but it is important in digestion. Complete the mini-crossword about bile.

   Across
   4. the acid in the stomach that bile will neutralise (12)
   5. what bile does to fat to break it into small droplets with a larger surface for enzyme action (10)
   6. where bile is stored (4, 7)

   Down
   1. where bile is made (5)
   2. describes the pH of bile (8)
   3. the enzyme in the small intestine that works faster in the alkaline conditions provided by bile (6)

BIOLOGY HIGHER

Main activity: Rates of enzyme reactions

Time 15 mins

Learning objectives
- To explain the effect of temperature and pH on the rate of reaction of enzymes

Equipment
- pencil

1. Enzymes have an optimum pH at which the rate of reaction is fastest. On the axes below sketch what the graph might look like for the enzymes a) protease in the stomach; b) salivary amylase. When you have finished, explain to your tutor why the lines follow the pattern you have drawn.

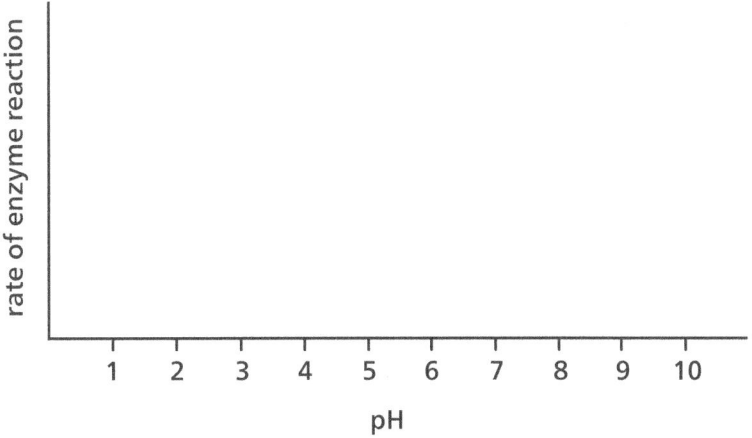

2. Enzyme reactivity is also affected by temperature.

   a) On the axes below, sketch what you would expect the graph would look like for a human digestive enzyme.

   b) When you have finished, label the optimum temperature.

   c) Annotate your graph to explain the shape of the line. Use the following keywords where relevant: kinetic energy, collisions, denature, active site, shape and substrate.

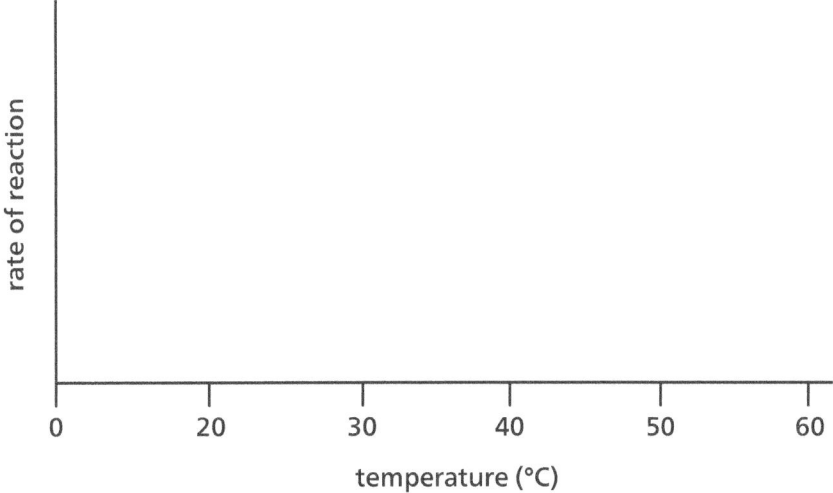

BIOLOGY HIGHER

Homework activity: Amylase investigation

Time 15 mins

Learning objectives
- To describe how qualitative reagents are used to test for food substances

Equipment
none

1. **A student carried out an investigation of the effect of pH on the rate of reaction of amylase enzyme. This is the plan that he made:**
   - Set up a spotting tile with a drop of iodine in each well.
   - Place 10 cm³ of starch solution into three boiling tubes.
   - Put the boiling tubes into a water bath at 35 °C.
   - Add 1 cm³ of buffer solution to each tube: either pH 3 buffer, pH 6 buffer or pH 9 buffer.
   - Add 1 cm³ of amylase solution to each tube.
   - Start the stop clock.
   - Every minute place two drops from each boiling tube onto the spotting tile, as shown in the diagram.
   - Record the colour of the iodine in each well.

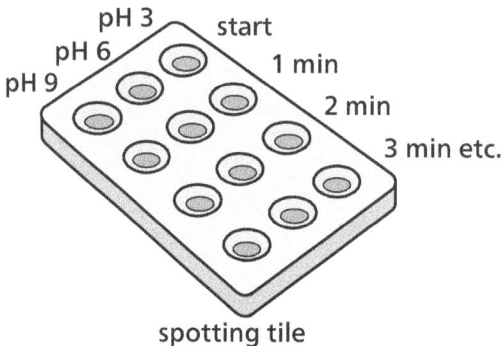

a) What was the purpose of the water bath?

_____

_____

b) What reaction is brought about by the amylase?

_____

_____

c) What is the purpose of the iodine solution?

_____

_____

d) The rate of reaction was greatest in the boiling tube at pH 6. The student reported that the investigation showed that the optimum pH for amylase is pH 6. Comment on the validity of this statement.

_____

_____

7 Answers

Starter activity: Organisation and the digestive system

1. Cell: B, tissue: A, organ: E, organ system: C, organism: D
2. Ensure student's answer includes: oesophagus or gullet, stomach, small intestine, large intestine, rectum, anus

Main activity: Enzymes and digestion

1. The student should draw a diagram of the substrate and enzyme joined as an enzyme–substrate complex, and a second diagram of the enzyme and products (now two separate parts).
2. Answers going across then down:
   Starch → simple sugars; used in respiration and to build new carbohydrates
   Stomach, pancreas and small intestine; to build new proteins
   Lipids → fatty acid + glycerol
3. Across: 4 hydrochloric; 5 emulsifies; 6 gallbladder
   Down: 1 liver; 2 alkaline; 3 lipase

Main activity: Rates of enzyme reactions

1. Protease should peak at a clearly acidic pH (pH 2). Salivary amylase should peak at a roughly neutral pH (pH 6–7). Both should fall to a reactivity of zero either side of this pH (although protease may simply have lower activity at pH 1).
2. a) The student should draw a curve with a steady increase in rate to a peak at around 40 °C (as this is a human enzyme), then a steeper decline to zero.
   b) The peak should be labelled 'optimum temperature'.
   c) Increasing the temperature and the kinetic energy of particles results in a greater number of effective collisions up to the optimum temperature. Above the optimum temperature the enzyme denatures; the shape of the active site changes, so the substrate can no longer bind. Increasing the temperature above the optimum denatures more and more enzyme molecules.

Homework activity: Amylase investigation

1. a) To control/maintain a constant temperature; to maintain the temperature at which the reaction will take place quickly
   b) It digests/breaks down starch into sugar
   c) To show when the reaction is complete/when the end point is reached; iodine turns from brown to blue-black in the presence of starch; when the starch has been broken down by the amylase it will remain brown
   d) It is not valid/the optimum pH cannot be determined accurately from this investigation; a greater number of different pH levels are needed/the optimum could be anywhere between pH 3 and pH 9

BIOLOGY HIGHER

8 Organisation: The heart, blood vessels and blood

Learning objectives

- To understand the structure and functioning of the human heart and lungs
- To understand the position and role of the heart's pacemaker cells
- To explain how the lungs are adapted for gaseous exchange
- To explain how the structure of blood vessels relates to their function
- To understand the double circulatory system and blood flow
- To understand the function of blood cells, platelets and plasma

Specification links

- 4.2.2.2
- 4.2.2.3

Starter activity

- **Heart and lungs structure; 10 minutes; page 60**

  The student needs to know the names (and functions) of the heart chambers and vessels. They should recognise the valves, but do not need to name each one. For the lungs, only the labels in bold are key knowledge. Remind the student about the plurals of terms such as atrium, alveolus and so on.

Main activities

- **Double circulation; 15 minutes; page 61**

  The boxes in question 1 should be provided to the student as shuffled cut-out cards and the instructions for a) and b) given verbally. The questions in part two are intended to be done verbally, using the heart diagram from the starter, but could be written if time permits. It is worth mentioning that double circulation is seen in mammals, but some other vertebrates, such as amphibians, have different circulation patterns.

- **Alveoli and blood vessels; 10 minutes; page 62**

  Ask the student to list the adaptations of the lungs for gaseous exchange as bullet points or as labels on the diagram. The student should be able to explain how the adaptations enable efficient gas exchange. The diagrams should show the thick wall and thick layer of muscle/elastic tissue in the artery compared with the vein. Make sure that the student does not have the misconception that this muscle contracts to pump blood along.

- **Blood as a tissue; 10 minutes**

  Ask the student to divide a piece of A4 paper into four quarters labelled: red blood cell, white blood cell, platelet and plasma. Then give them five minutes to record everything they know about each, then discuss their answers.

Plenary activity

- **Picture this; 5 minutes**

  Ask the student to choose any keyword from the lesson and draw a picture to express it. The tutor must guess what the word is and can then ask one question in relation to it. Repeat. This should be a quick-fire activity to check understanding.

Homework activity

- **Circulatory system quick quiz; 15 minutes; page 63**

  This is a short-answer quiz for the student to complete at home.

Support ideas

- **Double circulation** If the sorting activity proves difficult, use a red pencil for oxygenated and blue for deoxygenated blood, add arrows to the starter diagram to show blood flow, then try the activity again.
- **Alveoli and blood vessels** Support the student's understanding of pressure differences in the vessels by asking them to feel their pulse at the neck or wrist – this is the pressure surge of blood passing through elastic arteries as the heart contracts.

Extension ideas

- **Double circulation** Ask the student to name any arteries that they can think of and the organ that they supply.
- **Alveoli and blood vessels** Ask student to draw arrows onto the alveolus diagram to show diffusion pathways.

Progress and observations

BIOLOGY HIGHER

Starter activity: Heart and lungs structure   Time 10 mins

Learning objectives
- To understand the structure of the human heart and lungs

Equipment
- pencil

1. A simplified diagram of the heart is shown below. On this diagram, label the four chambers of the heart. Then add the names of the blood vessels that lead into and out of the heart. Finally, add as many other labels as you can.

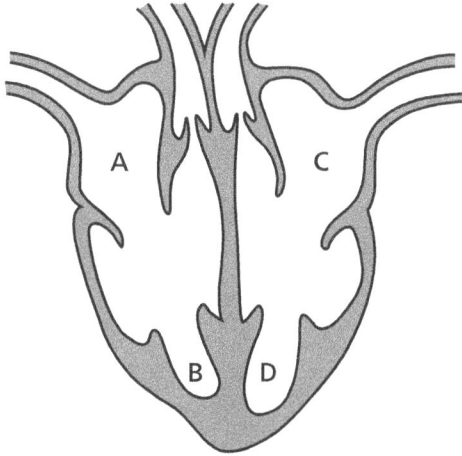

2. The coronary arteries are not shown on the diagram above. Why are the coronary arteries not shown?

_____

_____

3. The diagram below shows a pair of human lungs. One lung is cut away to show the inside. Add the labels in the table below to the diagram. Start with those in bold.

trachea	bronchus	alveolus	lung
rib	cartilage ring	pleural membrane	diaphragm

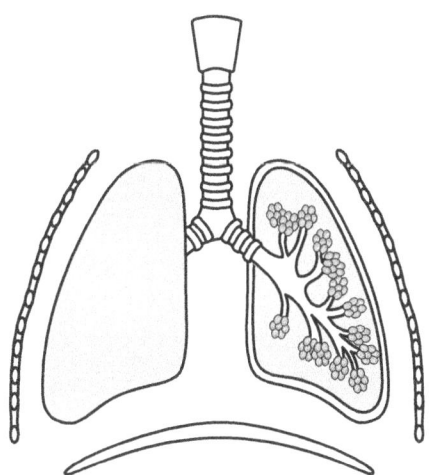

BIOLOGY HIGHER

Main activity: Double circulation

Time 15 mins

Learning objectives

- To describe the functioning of the human heart and circulatory system
- To understand the position and role of the heart's pacemaker cells

Equipment

- scissors

1. **Cut along the dotted lines around the boxes below until you have a set of cards. Shuffle them.**

   a) Place them in the correct order to show the flow of blood around one complete circuit of the double circulatory system. Start with the card in bold.

   b) Rearrange the cards so that events happening at the same time are placed side by side.

The vena cava brings deoxygenated blood from the body to the heart.	Blood enters the relaxed right side of the heart through the right atrium.	The right atrium contracts to finish filling the ventricle with blood.
The right ventricle contracts, squeezing the blood.	The valve between the atrium and ventricle snaps shut, preventing backflow to the right atrium.	The blood is pushed out of the heart through the pulmonary artery and to the lungs.
The right ventricle and right atrium relax.	Valves in the base of the pulmonary artery close to prevent backflow of blood into the right ventricle.	Blood starts to fill the right side of the heart again from the vena cava.
The pulmonary vein brings oxygenated blood from the lungs to the heart.	Blood enters the relaxed left side of the heart through the left atrium.	The left atrium contracts to finish filling the ventricle with blood.
The left ventricle contracts, squeezing the blood.	The valve between atrium and ventricle snaps shut, preventing backflow to the left atrium.	Blood is pushed out of the heart through the aorta and to the body.
The left ventricle and left atrium relax.	Valves in the base of the aorta close to prevent backflow of blood into the left ventricle.	Blood starts to fill the left side of the heart again from the pulmonary artery.

2. **The natural resting heart rate is controlled by a group of cells that act as a pacemaker.**

   a) How do they do this? _____

   _____

   b) Mark the position of these pacemaker cells on your diagram of the heart from the starter activity.

   c) What are artificial pacemakers? _____

   _____

BIOLOGY HIGHER

Main activity: Alveoli and blood vessels   Time 10 mins

Learning objectives
- To explain how the lungs are adapted for gaseous exchange
- To explain how the structure of blood vessels relates to their functions

Equipment
- *Heart and lungs structure* activity sheet

1. The diagram below shows an alveolus. The alveoli are the site of gas exchange in the lungs. Using this diagram and the diagram of the lungs from the starter activity sheet, list the adaptations of the lungs for gas exchange under the headings given.

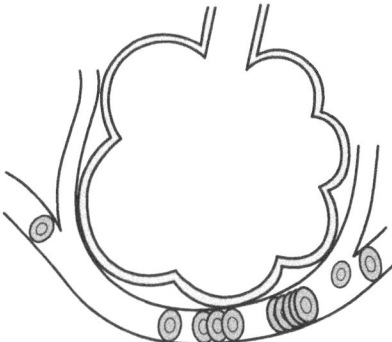

Gives large surface area for exchange _____

_____

Provides a short diffusion path _____

_____

Increases concentration gradient for diffusion _____

_____

2. Complete the table to summarise the adaptations of the blood vessels to their function.

	Artery	Capillary	Vein
Diagram			
Walls	has thick walls with lots of muscle and elastic tissue	thinnest walls, no muscle and elastic tissue	thin walls with little muscle and elastic tissue
Lumen	small	very small	large
Valves	no valves (except close to the heart)	no valves	has valves
Pressure	blood at high pressure	blood at lower pressure	blood at lowest pressure
Function	carry blood away from the heart	site of exchange of dissolved gases, food and waste products in tissues	carry blood back to the heart
How the blood vessel is adapted for its function			

62

 **BIOLOGY HIGHER**

Homework activity: Circulatory system quick quiz   Time **15 mins**

Learning objectives

- To understand the structure of the respiratory system
- To understand the double circulatory system and blood flow
- To understand the functions of blood cells, platelets and plasma

Equipment

none

1. Why is the circulation in mammals described as a double circulatory system? What is the advantage of having a double circulatory system?

2. The four main components of blood are: red blood cells, white blood cells, platelets and plasma. Name the component of the blood that would be linked to each of these statements.

   a) produces antibodies to fight off viral infections   _____

   b) carries oxygen from lungs to the tissues   _____

   c) carries digested food in solution   _____

   d) stick together to help form a blood clot   _____

   e) are disc-shaped cells with no nucleus   _____

   f) carries dissolved carbon dioxide   _____

   g) kill bacteria by engulfing and digesting them   _____

   h) are small 'fragments' of cells   _____

   i) contain the pigment haemoglobin   _____

3. The rate of blood flow (the volume pumped per minute) from the heart is called the cardiac output. This can be calculated in the following way:

   Cardiac output = heart rate (beats per minute) × stroke volume (volume pumped with each heartbeat)

   Work out the cardiac output for the following examples, include the correct units.

   a) A person who has a resting heart rate of 70 bpm and a stroke volume of $0.06 \, dm^3$. _____

   b) A runner who has a heart rate of 130 bpm with a stroke volume of $0.07 \, dm^3$. _____

4. Write these parts of the respiratory system in the order which oxygen is moved from the atmosphere to the blood.

        bronchus   capillary   alveolus   trachea   bronchiole

8 Answers

Starter activity: Heart and lungs structure

1. A: right atrium, B: right ventricle, C: left atrium, D: left ventricle
   Vessels left to right: vena cava, pulmonary artery, aorta, pulmonary vein
   Heart valves can just be labelled as 'valve'
2. The coronary arteries run over the outside of the heart to supply the heart muscle with blood. The diagram doesn't show the outside of the heart because it is a cross section.
3. Labels in bold added in correct place, others are not essential knowledge

Main activity: Double circulation

1. a) Correct order is shown in the table reading from left to right, starting in the top left hand corner
   b) The top half of the table shows the events in the right side of the heart and the bottom half the concurrent events in the left side. The cards should be laid out with matching statements next to each other and in order, showing double circulation.
2. a) By sending out regular electrical signals that initiate contraction of heart muscle
   b) Located in the right atrium
   c) Electrical devices used to correct irregularities in the heart rate

Main activity: Alveoli and blood vessels

1. Gives large surface area for exchange: there are many alveoli, and the folded surface of alveoli; provides a short diffusion path: alveolar wall and capillary wall both only one cell thick; increases concentration gradient for diffusion: good blood supply from many capillaries, ventilation by breathing
2. Artery walls; give the strength and elasticity needed to resist the high pressure and pulsing of blood as it is pumped by the heart; this maintains high pressure and keeps blood flowing quickly. Capillary walls: are thin and have a narrow lumen which brings the blood into close contact with cells and maximises diffusion and exchange. The pressure has reduced so there is no need for thick walls. Veins: have a wide lumen which provides less resistance so helps blood to flow at low pressure. The valves are needed to prevent backflow.

Homework activity: Circulatory system quick quiz

1. The blood travels through the heart twice in each complete circuit around the body. It keeps the pressure of the blood high and increases blood flow rates/efficiency of blood flow.
2. a) white blood cells
   b) red blood cells
   c) plasma
   d) platelets
   e) red blood cells
   f) plasma (RBC can carry CO_2 in a different form)
   g) white blood cells
   h) platelets
   i) red blood cells
3. a) $4.2\,dm^3\,min^{-1}$
   b) $9.1\,dm^3\,min^{-1}$
4. Trachea, bronchus, bronchiole, alveolus, capillary

9 Organisation: Non-communicable diseases

Learning objectives

- To evaluate treatments for coronary heart disease
- To describe the interactions between different types of disease
- To describe the effect of lifestyle on some non-communicable diseases
- To describe cancer and the nature of risk factors
- To interpret data about risk factors

Specification links

- 4.2.2.4, 4.2.2.5, 4.2.2.6, 4.2.2.7
- MS 4a

Starter activity

- **Types of disease; 5 minutes; page 66**

  Ask the student to complete the questions which explores their understanding of basic knowledge underlying this topic. Introduce health as the state of physical and mental well-being. Students should understand that lifestyle factors are personal behaviours or choices that may have an impact on health.

Main activities

- **Treatments and health issues; 15 minutes; page 67**

  Give the student time to complete the highlighting activity in question 1, then discuss the command term 'evaluate', which means the student must give an answer that balances both sides of the argument and use any information provided.

- **Lifestyle and cancer; 15 minutes; page 68**

  Question 1 will take more time than question 2. Prompts may be needed to encourage the student to remember facts, but make sure that enough thinking time is provided.

- **Sampling and correlation; 10 minutes**

  Ask the student to list important considerations when sampling to collect epidemiological data to ensure data is valid; for example, large sample size, representative samples (gender, age, socio-economic, geographical), avoiding bias. Ask the student to sketch what a scatter-graph might look like for body mass index against incidence of heart disease, and for amount of exercise against incidence of heart disease. The student should recognise positive and negative correlation. Discuss why finding a correlation between a factor and a disease does not necessarily mean that the risk factor causes the disease. A causal mechanism must be proved.

Plenary activity

- **Two stars and a wish; 5 minutes**

  Ask the student to tell you two things that they have learned in the session and one thing that they still wish to know more about. Take time to review this topic.

Homework activity

- **Health studies; 15 minutes; page 69**

  This homework uses data from an epidemiological study to develop data interpretation skills.

Support ideas

- **Treatments and health issues** Show diagrams from medical health websites to illustrate stents and artificial hearts/valves.
- **Lifestyle and cancer** Draw or show a series of simple diagrams to show how malignant tumours lead to secondary tumours.

Extension ideas

- **Treatments and health issues** Ask the student to decide whether they would choose biological or mechanical heart valve treatment and ask them to explain why.
- **Lifestyle and cancer** Ask the student to suggest links between the topic of cancer and that of cell division. Look for consideration of mitosis and the cell cycle, checks of DNA, along with mutations of genes controlling the cell cycle.

Progress and observations

Starter activity: Types of disease   Time 5 mins

Learning objectives
- To identify non-communicable diseases
- To recall some lifestyle and environmental risk factors
- To explain the symptoms of coronary heart disease

Equipment
none

1. A non-communicable disease is a medical condition which is not passed on from person to person. That is, they are not caused by infectious microorganisms. Circle the non-communicable diseases in the list below.

   coronary heart disease   chicken pox
   influenza   stroke
   measles   skin cancer
   asthma   polio
   salmonella   diabetes

2. Risk factors are linked to an increased rate of a disease. They can be lifestyle factors or substances in the person's body or environment. Name one lifestyle factor and one substance that are risk factors for non-communicable diseases along with the name of the disease that they are linked to.

   _____
   _____

3. In coronary heart disease, layers of fatty material build up inside the coronary arteries.

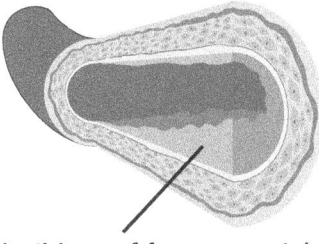

build-up of fatty material

   a) What are the symptoms of coronary heart disease?
   _____
   _____

   b) Why do the fatty deposits cause these symptoms?
   _____
   _____
   _____

BIOLOGY HIGHER

Main activity: Treatments and health issues

Time 15 mins

Learning objectives
- To evaluate treatments for coronary heart disease
- To describe the interactions between different types of disease

Equipment
- two highlighter pens of different colours

1. There are a variety of treatments for cardiovascular disease. They have benefits but also carry risks. The passage below is about some of these treatments. Read the passage and circle the names of any treatments. Then highlight the benefits of treatments in one colour and the risks in another colour.

> Stents are a mechanical devices that can keep the coronary arteries open when they have become blocked. They consist of an expandable metal mesh and are inserted through a blood vessel using keyhole surgery.
>
> Patients do not need general anaesthetic and usually return to normal activities quickly. They will need to take medication for life to prevent clots from forming. Complications are rare but may be severe, such as a heart attack.
>
> Statins are a class of drugs that are very effective at lowering cholesterol levels in the blood. This slows down the rate of fatty material deposit in the coronary arteries, which in turn reduces the risk of heart disease. Statins stop cholesterol from being made in the liver. Statins can cause muscle pain or digestive problems in some people. In rare cases they may cause liver damage. Most people taking statins do not experience side effects.
>
> Sometimes heart valves may become faulty. The valve may not open properly, restricting the flow of blood, so the heart must pump harder to force the blood through. Heart valves can also develop a leak, allowing blood to flow backwards, so the heart must work harder to pump the same volume of blood. Faulty heart valves can be replaced using biological or mechanical valves. Both are very effective treatments. Mechanical valves are made of artificial material, which increases the risk of a blood clot developing, so anti-clotting drugs must be taken for life after the operation. Biological valves are made from animal tissue, usually from pigs. There is less risk of clots but biological valves do not last as long as mechanical ones and may have to be replaced. For both treatments there is a risk of infection following the major operation.
>
> In the case of heart failure, a donor heart can be transplanted. This can extend the life of patients by many years, but there may be a long wait for a donor. There is a risk of transplant rejection, which can be fatal. Patients must take drugs to suppress their immune system, but these have side effects. Artificial hearts are readily available and may occasionally be used to keep patients alive whilst waiting for a heart transplant, or to allow the heart to rest as an aid to recovery. Artificial hearts may be faulty and carry a risk of blood clots or infection following surgery.

2. Different types of disease may interact. Join the boxes to show some of these interactions.

defects in the immune system	can trigger allergies such as skin rashes and asthma
viruses living in cells	can lead to depression and other mental illness
immune reactions caused by a pathogen	more likely to suffer from infectious diseases
severe physical ill health	can be the trigger for cancers

BIOLOGY HIGHER

Main activity: Lifestyle and cancer   Time 15 mins

Learning objectives
- To describe the effect of lifestyle on some non-communicable diseases
- To describe cancer and the nature of risk factors

Equipment
none

1. Risk factors may be linked to an increased rate of a disease. Read each of the following statements about risk factors, then tell your tutor three additional facts about each.

   a) Diet, smoking and exercise are linked to cardiovascular disease.

   b) Obesity is a risk factor for Type 2 diabetes.

   c) Drinking alcohol can affect liver and brain function.

   d) Smoking is linked to lung disease and lung cancer.

   e) Smoking and drinking alcohol when pregnant can harm unborn babies.

   f) Carcinogens, including ionising radiation and some chemicals, are risk factors for cancer.

2. Cancer occurs when changes in cells cause uncontrolled growth and division leading to malignant tumours. Benign tumours are not cancerous. Complete the table to summarise the difference between the tumour types.

Type of tumour	Characteristics
Benign	_____ growth
	Usually surrounded by a _____, so they can be removed easily
	Rarely spread to other parts of the body
	They can press on other body organs and look unsightly
Malignant	Grow faster
	Cancer cells detach and move into the bloodstream
	Cancer cells carried in the blood can form _____ tumours in other parts of the body
	_____ growth prevents organs from working properly

BIOLOGY HIGHER

Homework activity: Health studies   Time 15 mins

Learning objectives
- To interpret data about risk factors

Equipment
none

Health is a state of physical and mental well-being. Diseases are major causes of ill health, but other factors such as diet, stress and life situations can have a major effect on both physical and mental health.

The graph shows the results of a study on the effect of stress on mental health. Interviews were carried out on 1656 adults who did not have a mental health condition.
They were visited again after 18 months and the percentage of men and women who had now developed a mental disorder was assessed. The number of stressful life events that had happened to them over the 18 months was also recorded.

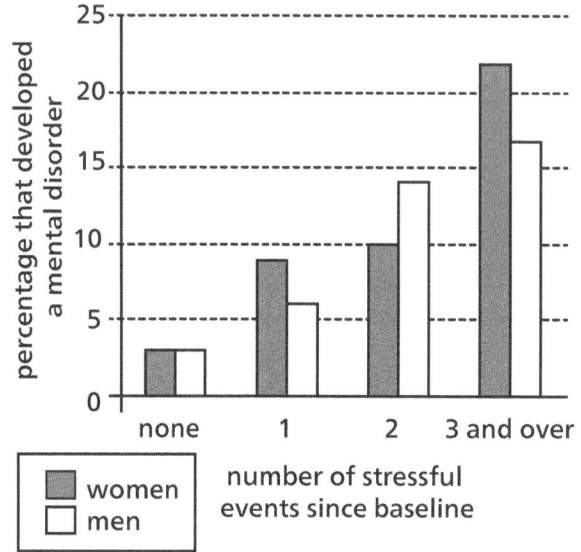

Use information from the graph to answer the following questions.

1. Describe the effect of stressful life events on the mental health of men.

   _____
   _____
   _____
   _____

2. Compare the effect of stressful life events on men and women.

   _____
   _____
   _____

3. We cannot conclude from this study that stress causes mental health conditions. Explain why.

   _____
   _____

BIOLOGY HIGHER

9 Answers

Starter activity: Types of disease

1. Coronary heart disease, stroke, skin cancer, asthma, diabetes
2. Any valid response relating to non-communicable disease, such as lifestyle: overeating/lack of exercise linked to heart disease; smoking linked to lung cancer
   Substances: car exhaust pollution linked to asthma; pollutants in food linked to cancer
3. a) Chest pain, shortness of breath
   b) The deposits cause a narrowing of the artery. This reduces the flow of blood through the coronary arteries, resulting in a lack of oxygen for the heart. The heart muscle can no longer work efficiently.

Main activity: Treatments and health issues

1. Stents: benefits include keyhole surgery, return to normal activities quickly, complications are rare; risks include the need to take medication for life, clots, heart attack
   Statins: benefits are they're very effective, reduce the risk of heart disease; risks include muscle pain, digestive problems, liver damage
   Biological valves: benefits are they're very effective; risks: do not last as long as mechanical ones, risk of infection, major operation
   Mechanical valves: benefits are they're very effective; risks include blood clots, drugs must be taken for life, risk of infection, major operation
   Heart transplant: benefits include extension of life; risks are the long wait for donor, rejection can be fatal, drug side effects Artificial hearts: benefits are they're readily available; risks are they can be faulty, risk of blood clots or infection
2. Defects in the immune system; more likely to suffer from infectious diseases
   Viruses living in cells; can be the trigger for cancers
   Immune reactions caused by a pathogen; can trigger allergies such as skin rashes and asthma
   Severe physical ill health; can lead to depression and other mental illness

Main activity: Lifestyle and cancer

1. Key facts might include the following:
   a) Links to obesity, the raising of blood pressure and damage to arteries, increased cholesterol and deposit of fatty material
   b) Cells become resistant to insulin, they no longer respond to remove glucose from the blood, excess blood sugar damages cells, diabetes can raise blood pressure and add to the risk of cardiovascular disease
   c) Short-term and permanent effects on behaviour, memory and coordination, cirrhosis of liver, death of cells, fatty deposits
   d) Damage to epithelium, paralysis of cilia, mucus clearing, emphysema, infections/bronchitis, chemicals damage DNA leading to cancer
   e) Smoking: greater risk of miscarriage, birth defects, sudden infant death syndrome, low birth weight
      Alcohol: learning difficulties, developmental problems, foetal alcohol syndrome
   f) Damage to DNA, uncontrolled cell division, pollutants, UV radiation and skin cancer
2. Slow, membrane, blood, secondary, tumour

Homework activity: Health studies

1. Percentage of men with mental health conditions gradually increases with increasing number of stressful life events; between 3% and 8% increase for each additional stressful event; biggest increase of 8% is between one and two stressful events; 3% of men developed a mental health condition even without any stressful event
2. Women show the same pattern of increased incidence of mental health conditions with increased stress; women are more affected by the first and third event/men more affected by second event; figures from graph used to support this argument
3. Although there appears to be a correlation we would need to prove the causal mechanism; there might be another linked factor; poor mental health might make stressful life events more likely to happen; mental health conditions developed in 3% of men and women even without any stressful event

10 Organisation: Plant tissues, organs and systems

Learning objectives

- To explain how the structure of plant tissues relates to their function
- To describe the structure of the leaf as a plant organ
- To describe the process of translocation and transpiration
- To explain the effect of environment on transpiration rate

Specification links

- 4.2.3.1
- 4.2.3.2
- MS 2b

Starter activity

- **Plant tissues; 5 minutes; page 72**

  An introduction to plant tissues. Complete the table together, letting the student take the lead; explain any misconceptions. Show the student a house plant and talk about where each of the plant tissues are found.

Main activities

- **Plant organs and transport systems; 15 minutes; page 73**

  Do not give the student the activity sheet immediately. Show the student the leaf diagram for 30 seconds. Then allow three minutes for the student to draw what they can remember. Then give a further two minutes to add as many labels as they can. Cut the dominoes along the solid lines. They have the first half of one sentence and the second half of another sentence separated by a dotted line. Starting with the statement in bold, the student should arrange the dominoes in a line so that the sentences give a correct description. The final word is also in bold. Use the activity as a focus for discussion of the processes described.

- **Transpiration; 15 minutes; page 74**

  In question 1, check that the student understands the role and function of the guard cells and stomata. Question 2 provides the opportunity to check understanding of transpiration investigations and the factors that affect transpiration rate.

  The questions can be answered verbally with annotations made if required.

- **Cell adaptations; 10 minutes**

  The student should be able to explain how the structure of root hair cells, xylem and phloem are adapted to their functions. Ask the student to sketch each cell and explain the adaptations. Link the root hair cell and xylem to the transpiration stream and phloem to translocation.

Plenary activity

- **Gas exchange and evaporation; 5 minutes**

  Ask the student to draw arrows in different colours to show the movement of CO_2, oxygen and water in bright sunlight (when the plant is photosynthesising and stomata are open). Then ask: how would this differ in the dark? Emphasise that plants do respire. Gas exchange in relation to photosynthesis will be covered later.

Homework activity

- **Stomata investigation; 15 minutes; page 75**

  The homework focuses on interpretation of data and maths skills in relation to investigations of stomata.

Support ideas

- **Plant tissues** The student may be uncertain about the shape and appearance of cells in these tissues. Show the student one of the many photomicrographs available online (search 'leaf transverse section').
- **Transpiration** To support understanding of how guard cells open when turgid, blow up a long balloon that has sticky tape attached along one side. This thickening cannot stretch as much, so the balloon will curl when full of air.

Extension ideas

- **Plant organs and transport systems** Ask the student to label the tissues in a simple diagram of a section through a stem.
- **Transpiration** Ask the student to sketch the pathway of water and minerals into a plant.

Progress and observations

Starter activity: Plant tissues

Time 5 mins

Learning objectives
- To explain how the structure of plant tissues relates to their function

Equipment
- house plant

1. Complete the boxes to show the structure of some plant tissues and their associated functions.

Tissue	Structure	Function
	hollow tubes strengthened by lignin	
Phloem tissue	tubes of elongated cells, with pores in the end walls which allow cell sap to move between cells	
	a single outer layer of flattened cells, transparent, no chloroplasts; may be covered in a waxy cuticle	a protective layer that covers the whole plant; reduces and controls water loss.
Palisade mesophyll	cell shape allows them to be tightly packed; cells contain lots of chloroplasts	
	cells are loosely packed with a large surface area and have fewer chloroplasts	the main site of gas exchange; gases can diffuse easily between cells; some photosynthesis
Meristem tissue		found at growing tips of plants; cells can differentiate into a variety of different cell types

2. Look at a living plant. Explain to your tutor where you might find each type of tissue.

BIOLOGY HIGHER

Main activity: Plant organs and transport systems

Time 15 mins

Learning objectives
- To describe the structure of the leaf as a plant organ
- To describe the processes of translocation and transpiration

Equipment
- spare paper
- scissors

1. The diagram shows a section through a leaf. Your tutor will show you this diagram for 30 seconds. You then have three minutes to draw what you remember. You will then have a further two minutes to add as many labels as you can.

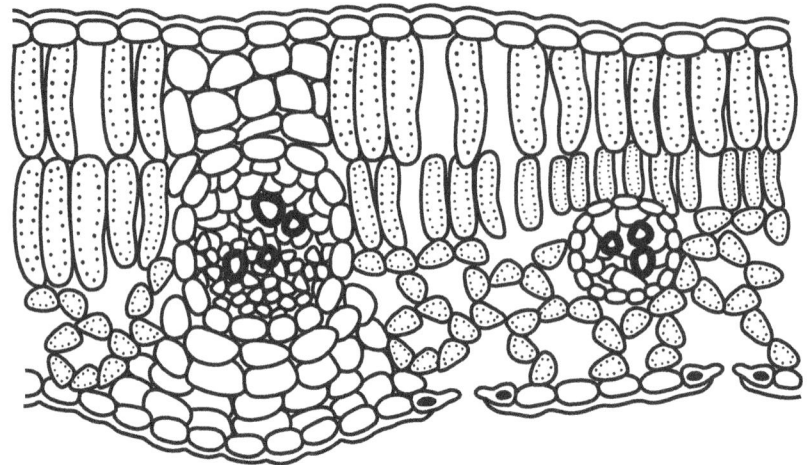

2. Cut along the dotted lines around the boxes so you have a set of 'dominoes' that have information about the plant transport system. Start with the statement in bold, then arrange the dominoes in a line so that the sentences give a correct description. The final word is also in bold.

	by osmosis, and mineral ions by active transport.	or close the stomata.
The roots, stem and leaves form a	The water and minerals are then transported in a tissue called	In this way guard cells control the
plant organ system for transport of substances around the plant	xylem to the stems and leaves.	rate of gas exchange and water loss.
Transpiration is the loss of water from a plant by the process of	Water may be used by the plant, but most evaporates through	Phloem tissue transports dissolved sugars from
evaporation from the leaves.	pores called stomata in the leaves.	the leaves to the rest of the plant.
This causes water movement from root to leaf known as the	The stomata are surrounded on each side by cells called	The sugar may be used immediately, for example in
transpiration stream.	guard cells.	respiration, or it may be stored as starch for later use.
Root hair cells are adapted for the efficient uptake of water	These can change shape to open	The movement of food molecules through phloem tissue is called **translocation.**

BIOLOGY HIGHER

Main activity: Transpiration

Time 15 mins

Learning objectives
- To describe the function of the stomata and guard cells
- To explain the effect of environment on transpiration rate

Equipment
- pencil
- spare paper

1. The role of stomata and guard cells are to control gas exchange and water loss.

a) Complete the sentences below.

Open stoma	Closed stoma
When guard cells take up water by _____ they become turgid. The thicker inner cell walls become curved and the stoma pore _____ . Stomata usually open during daylight to allow gas exchange for photosynthesis.	At night, water moves out of the guard cells by osmosis. The cells become _____ and the stoma pore closes. This reduces _____ loss at night when less gas exchange is needed. Stomata will also close during the day if the plant loses too much water and begins to wilt.

b) On a separate piece of paper, sketch a diagram to show what the guard cells and stoma would look like in each case.

2. The diagram below shows a piece of equipment called a potometer. This equipment can be used to measure the rate of transpiration. Explain to your tutor how this equipment works.

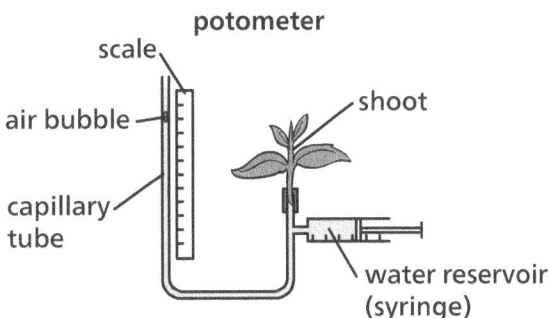

A student wants to investigate the effect of changing light intensity on the rate of transpiration. She controls the area of leaf by using the same plant shoot for each experiment.

a) What other factors must she control that would affect the rate of transpiration?

_____

b) Explain to your tutor how each factor, including light, affects transpiration.

c) The scale measures water uptake in mm^3. In the first measurement 15 mm^3 of water is taken up in 10 minutes. What is the rate of water uptake? Is this the same as the rate of transpiration?

_____

d) What other method could be used to measure transpiration rate? _____

BIOLOGY HIGHER

Homework activity: Stomata investigation

Time: 15 mins

Learning objectives
- To process data from investigations involving stomata

Equipment
- calculator

1. The density of stomata were investigated on the leaves of two house plants. Stomata were counted in 0.5 mm² areas sampled from the underside of five leaves from each plant.

	Number of stomata counted in a 0.5 mm² area					Mean count	Mean density of stomata mm^{-2}
plant A	121	133	112	124	116		
plant B	22	32	25	23	26		

a) Calculate the mean number of stomata counted in a 0.5 mm² area and add this to the table.

b) Use these means to calculate the mean stomata density per square mm.

c) Why were stomata counted from five leaves for each plant?

d) Why were all samples taken from the underside of leaves?

e) Which plant is more likely to have adaptations for living in dry conditions and how do you know?

10 Answers

Starter activity: Plant tissues

1.

Tissue	Structure	Function
Xylem tissue	Hollow tubes strengthened by lignin	Transports water and mineral ions from the roots to the stems and leaves
Phloem tissue	Tubes of elongated cells; pores in the end walls allows cell sap to move between cells	Transports dissolved sugars from the leaves to the rest of the plant
Epidermal tissue	A single outer layer of flattened cells; transparent, no chloroplasts; may be covered in a waxy cuticle	A protective layer that covers the whole plant; reduces and controls water loss
Palisade mesophyll	Cell shape allows them to be tightly packed; cells contain lots of chloroplasts	The main site of photosynthesis
Spongy mesophyll	Cells are loosely packed with a large surface area and have fewer chloroplasts	The main site of gas exchange; gases can diffuse easily between cells; some photosynthesis
Meristem tissue	Contains unspecialised and rapidly dividing cells	Found at growing tips of plants; cells can differentiate into a variety of different cell types

2. Key points are: that xylem and phloem connect root tips to leaf tips, palisade and spongy mesophyll are associated with the leaf, meristem at growing tips, epidermis is the outer layer of whole plant, there are different types on upper and lower leaf

Main activity: Plant organs and transport systems

1. Labels should include epidermis, palisade and spongy mesophyll, xylem and phloem, and guard cells surrounding stomata
2. Correct order is shown on activity sheet, answers going down then across

Main activity: Transpiration

1. a) Open stoma: osmosis; opens. Closed stoma: flaccid; water
   b) Guard cells should be turgid and bent with the stomata open; flaccid, straight guard cells and the stomata closed
2. a) Temperature, humidity, air movement
   b) An increase in temperature increases the kinetic energy of water vapour molecules and increases the rate of diffusion and evaporation. An increase in humidity reduces the concentration gradient and reduces diffusion and evaporation rates. Increased air movement increases the concentration gradient and increases diffusion and evaporation rates. Increased light causes more photosynthesis and more stomata to open, so greater water loss would occur.
   c) 1.5 mm^3 min^{-1}. It is not quite the same because a small amount water may be used by the plant, for example in photosynthesis, but in this experiment, it is assumed to be the same.
   d) The change in mass of a plant could be used. Mass is lost as water is lost through transpiration.

Homework activity: Stomata investigation

1. a) Plant A: 121, plant B: 26
   b) Plant A: 242, plant B: 51
   c) To gain a more representative sample/achieve a more representative mean
   d) Because in most plants there are very few stomata on the upper side of the leaf
   e) Plant B, because it has a reduced number of stomata which would help to conserve water

11 Infection and response: Communicable diseases – viral and bacterial diseases

Learning objectives

- To understand which organisms are pathogens and how diseases are spread
- To explain how the spread of diseases can be reduced or prevented
- To describe the symptoms, spread and control of some bacterial and viral diseases

Specification links

- 4.3.1.1
- 4.3.1.2
- 4.3.1.3

Starter activity

- **Infectious diseases crossword; 5 minutes; page 78**

  Ask the student to complete the crossword. Discuss any key terms they are uncertain about.

Main activities

- **Bacteria and viruses as pathogens; 15 minutes; page 79**

  For question 1 a) discuss the diagram of HIV reproduction then ask the student to complete the flow diagram.

  Details of RNA reverse transcription are not needed – the idea of reproduction inside host cells and the damage done is key. Questions b) – e) could be answered verbally with notes made on the sheet if required.

- **Examples of bacterial and viral diseases; 20 minutes; page 80**

  These diseases are required examples from the specification. Discuss the diseases one at a time, asking the student to write what they know about each disease in terms of spread, symptoms and control. Then ask prompt questions and assist with completion of the spider diagram. The information provided on the answer sheet mostly reflects specification content.

- **Compare and contrast bacteria and viruses; 5 minutes**

  Remind the student of the structure of a bacterium as covered in the first lesson. Ask students to list any similarities and differences in structure and life cycle. Sketch a virus or show pictures if the student is uncertain.

Plenary activity

- **Word association; 5 minutes**

  Say a keyword from the activity sheet and the student must immediately reply with an associated word, then repeat.

  This should be a quick-fire activity. The five example diseases are a good place to start. Stop to review any topics for which the student has limited recall.

Homework activity

- **Communicable disease questions; 15 minutes; page 81**

  Ask the student to complete the short exam-style questions, writing on the sheet.

Support ideas

- **Bacteria and viruses as pathogens** Show pictures on the internet of the four different types of pathogens if the student is uncertain about the appearance of each, but note that this will be covered further in the next lesson.
- **Examples of bacterial and viral diseases** Showing the student photos, especially of the effects of TMV and measles, will help them to remember the disease symptoms.

Extension ideas

- **Bacteria and viruses as pathogens** Ask the student to name diseases that are spread via water. Discuss ways in which diseases spread by direct contact, aerosol and water can be reduced.
- **Examples of bacterial and viral diseases** Ask the student to explain how vaccinating young children against measles helps to reduce its spread through a population when there is a measles outbreak.

Progress and observations

Starter activity: Infectious diseases crossword

Time 5 mins

Learning objectives
- To understand what types of organisms are pathogens and how diseases are spread

Equipment
- pencil

1. Complete the crossword about infectious diseases.

Across
3. one of the three pathways by which disease can be spread (5)
6. microorganisms that cause infectious disease (9)

Down
1. live and reproduce inside cells, causing cell damage (7)
2. produced by sneezing, they spread disease through the air (8)
4. illness that may be caused by viruses, bacteria, protists or fungi (7)
5. another word for an infectious disease (12)
7. produced by bacteria, they cause tissue damage and make us feel ill (6)

BIOLOGY HIGHER

Main activity: Bacteria and viruses as pathogens

Time 15 mins

Learning objectives

- To understand how bacteria and viruses cause the symptoms of disease

Equipment

none

1. Both bacteria and viruses may reproduce rapidly inside the body. Viruses cannot live in isolation and must infect body cells to reproduce. HIV is a virus, the diagram shows how HIV reproduces inside cells.

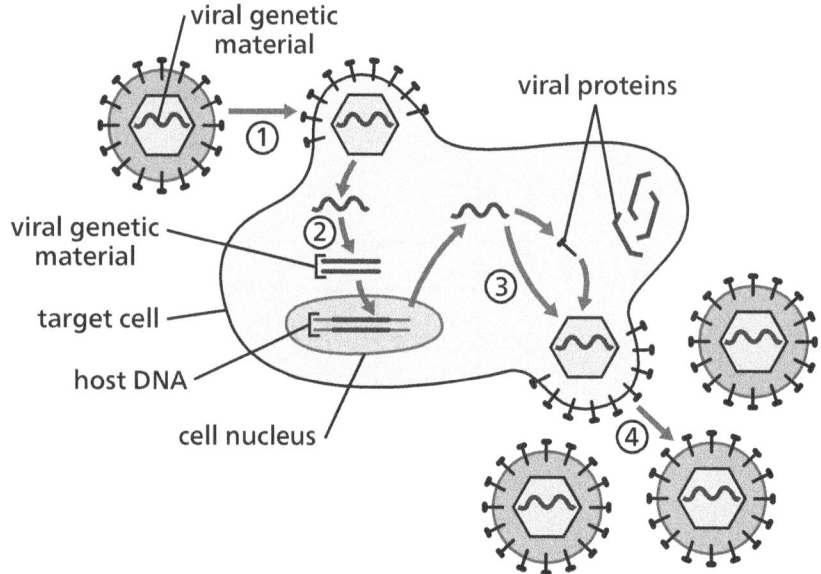

a) Discuss the diagram with your tutor, then produce a flow diagram to describe what is happening in stages 1 to 4.

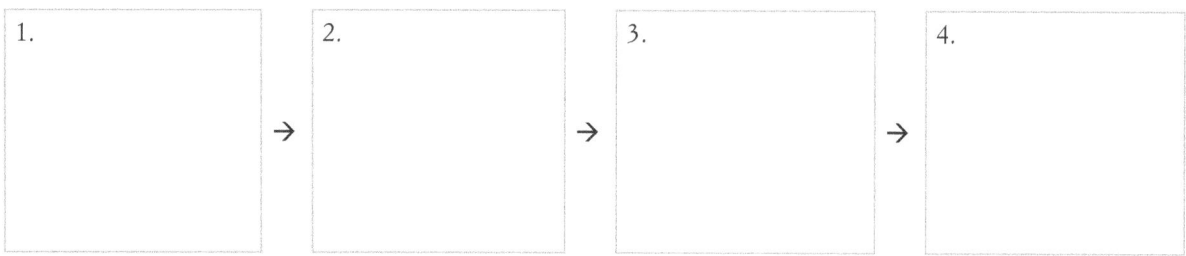

b) What is the main type of cell that HIV infects? _____

c) Explain how the process of reproduction inside cells eventually causes the symptoms of AIDS.
_____

d) Most bacteria reproduce in body fluids rather than inside body cells. How do bacteria reproduce?
_____

e) How do bacteria cause damage to cells? _____

BIOLOGY HIGHER

Main activity: Examples of bacterial and viral diseases

Time 20 mins

Learning objectives

- To describe the symptoms, spread and control of some bacterial and viral diseases

Equipment

- two highlighters of different colours

1. On the diagram below, highlight diseases caused by bacteria in one colour and those caused by viruses in another. Then construct a spider diagram to describe each disease. You should include information about how the disease is spread, the symptoms and damage done by the pathogen, as well as the control or treatment methods.

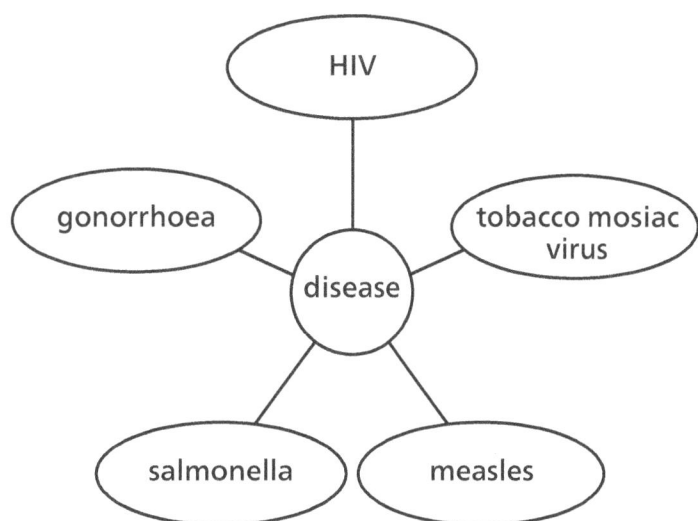

80

BIOLOGY HIGHER

Homework activity: Communicable disease questions

Time 15 mins

Learning objectives

- To describe the symptoms, spread and control of some bacterial and viral diseases

Equipment

none

1. Infectious diseases are caused by the spread of microorganisms. Describe three ways that infectious diseases may be spread.

2. Measles is a disease caused by a virus. Explain how the spread of measles can be reduced.

3. Salmonella is a disease that is caused by a type of bacteria. Describe the symptoms of salmonella.

4. Explain how initial infection with small numbers of salmonella bacteria can lead to these symptoms.

11 Answers

Starter activity: Infectious diseases crossword

Across: 3. water 6. pathogens; Down: 1. viruses 2. droplets 4. disease 5. communicable 7. toxins

Main activity: Bacteria and viruses as pathogens

1. a) 1 Virus particle enters cell. 2 Virus genetic material is inserted into host cell DNA. 3 New virus proteins and genetic material is made by the host cell. 4 New virus particles form and burst out of the cell, causing damage or killing the cell.
   b) White blood cells/cells of the immune system/T helper cells
   c) The immune system becomes so badly damaged that it can no longer fight off other infections or destroy cancer cells.
   d) Binary fission
   e) Produce toxins/poisons that damage body cells

Main activity: Examples of bacterial and viral diseases

1. Student's own answer. Check they have demonstrated the following key knowledge on spread, symptoms and control:
   Viruses: measles: spread in air, by inhalation of droplets from sneezes and coughs; fever, red skin rash, fatal if complications arise; vaccination of young children
   HIV: direct contact, by sexual contact or exchange of body fluids such as blood when drug users share needles; initially causes a flu-like illness, attacks the body's immune cells, AIDS arises when the body's immune system becomes seriously damaged, leading to secondary infections or cancers; controlled with antiretroviral drugs
   TMV: direct contact, such as from other plants or farm workers; affects many species of plants including tomatoes, 'mosaic' pattern of discolouration on leaves, reduces photosynthesis and hence plant growth is stunted; sanitation, selective breeding
   Bacteria: salmonella: direct contact, spread by bacteria ingested in food, contamination of food by faeces; fever, abdominal cramps, vomiting and diarrhoea, caused by the bacteria and the toxins they secrete; poultry are vaccinated against salmonella (in UK), prepare food in hygienic conditions, hand washing with soap
   Gonorrhoea: direct contact, sexually transmitted (STD); thick yellow/green discharge from the vagina or penis, pain on urinating; reduce spread/treat with antibiotic penicillin, problem with antibiotic resistant strains, use a barrier method of contraception

Homework activity: Communicable disease questions

1. By direct contact; by water; by air
2. Vaccination of young children; this reduces spread through the population as there are fewer people that can pass it on; any valid method of avoiding aerosol/airborne infection
3. Fever; abdominal cramps; vomiting; diarrhoea
4. Bacteria reproduce rapidly in the body/by binary fission; the bacteria invade the cells lining the gut which causes damage to these cells; they release toxins/poisons that cause damage to body cells

12 Infection and response: Communicable diseases – fungal and protist diseases

Learning objectives

- To describe the characteristics of fungi and protists
- To describe the life cycle of malaria
- To explain the symptoms, spread and control of malaria
- To explain the symptoms, spread and treatment of rose black spot disease

Specification links

- 4.3.1.4
- 4.3.1.5

Starter activity

- **Comparing pathogens; 5 minutes; page 84**

  Ask the student to complete the table to compare the four types of pathogen that are identified in the specification sections on communicable diseases.

Main activities

- **Rose black spot, a fungal disease; 15 minutes; page 85**

  Ask the student to read the 'magazine article'. Then go through the questions; this could be done verbally, or the student could write on the sheet.

- **Malaria, a protist disease; 15 minutes; page 86**

  Ask the student to put the labels in the correct place and then discuss the answers before sticking them on to a separate piece of paper. If time is short, numbers could be used to represent the label, but this does not provide as strong a visual reminder. Use the remaining questions to form the basis of a discussion to check understanding. The student should know that malarial parasites reproduce asexually in the human host, but sexually in the mosquito (mentioned also in lesson 25) – this could be added as a label on the diagram.

- **Preventing the spread of malaria; 10 minutes**

  Ask the student to list all the ways they can think of by which the spread of malaria can be controlled. This should include ways of preventing mosquitoes from breeding and using mosquito nets to avoid being bitten.

Plenary activity

- **Dubious statements; 5 minutes**

  Ask the student to comment on the following statements: 'plant diseases are less of a concern than diseases of humans'; and 'mosquitoes cause malaria'. Encourage students to talk about what might be true, if anything, about the statement (for example, no vectors means no malaria). They should also state what is false (for example, crop diseases vastly reduce food available to a growing human population, *Plasmodium* protist causes the symptoms of malaria).

Homework activity

- **Mosquitoes and malaria prevention; 15 minutes; page 87**

  Ask the student to answer the short exam-style questions on the activity sheet.

Support ideas

- **Rose black spot, a fungal disease** Show photos of rose leaves with black spot to support understanding of the effect of the disease on photosynthesis.
- **Malaria, a protist disease** Show a short online video about the cause and symptoms of malaria. A search for 'video – what is malaria' will give several useful results.

Extension ideas

- **Rose black spot, a fungal disease** Ask the student to tell you the names and symptoms of some human fungal diseases.
- **Malaria, a protist disease** Go into a little more detail about the life cycle of the malaria protist, in particular that there are many different stages of development, as well as an asexual and a sexual phase of reproduction.

Progress and observations

BIOLOGY HIGHER

Starter activity: Comparing pathogens Time 5 mins

Learning objectives
- To compare the characteristics of fungi, protists, bacteria and viruses

Equipment
none

1. There are a variety of groups of microorganisms. Within each group, there are some types that cause disease, while others may be useful. Complete the table below to summarise the characteristics of groups of microorganisms that can act as pathogens.

Microorganism group	Characteristics of microorganism	Example of pathogen
virus	They are not cells; they are only made up of a protein coat and genetic material. They do not feed or respire, but rely on host cells for _____. They cause damage to host cells when _____ particles burst out from cells.	
bacteria	_____ cells do not have a nucleus. Reproduce asexually by _____ fission. Pathogenic bacteria produce _____ which damage body cells.	
fungus	Eukaryotic cells contain sub-structures such as a n_____ . Single celled or multiple celled, include yeasts, m_____ and mushrooms. May _____ sexually or asexually using dispersal cells called spores.	rose black spot
protist	E_____ cells contain sub-structures such as a nucleus. Can be single celled or multiple celled. There are many different types. May reproduce sexually or asexually, often have complex life _____ .	

84

BIOLOGY HIGHER

Main activity: Rose black spot, a fungal disease

Time 15 mins

Learning objectives
- To explain the symptoms, spread and treatment of rose black spot disease

Equipment
none

1. Read the article about rose black spot disease, then answer the questions.

> **ROSE BLACK SPOT DISEASE – HOW TO BEAT IT!**
> **GARDENS AND FLOWERS MAGAZINE**
>
> Black spot is a serious disease affecting roses. It is caused by a fungus, *Diplocarpon rosae*, which infects the leaves and greatly reduces plant vigour.
>
> If you have seen purple or black patches appear on the upper leaf surfaces of your roses, then you might be dealing with rose black spot disease. You may even be able to see a few strands of the fungus (hyphae) growing around the edge of the spots.
>
> On severely affected roses the leaves will turn yellow and drop off early. The rose plants will not grow well and may become stunted.
>
> There are two main ways to deal with this disease. First, try non-chemical control. Cut off all infected leaves and collect fallen leaves, then destroy them. A good excuse for a bonfire!
>
> This treatment doesn't always work. If plants become re-infected, try chemical treatment. We recommend using 'Spot-Away' spray at the first sign of black spot.

a) Explain why rose black spot results in poor growth of the plant.

_____
_____
_____

b) Explain why removing and destroying affected leaves helps to control the disease.

_____

c) Explain how the disease may reappear, even when all infected leaves have been destroyed.

_____
_____

d) What type of chemical is likely to be found in 'Spot-Away' spray?

_____

BIOLOGY HIGHER

Main activity: Malaria, a protist disease

Time 15 mins

Learning objectives

- To describe the life cycle and explain the symptoms of malaria

Equipment

- scissors
- glue
- spare paper

1. **The pathogen that causes malaria is a type of protist called *plasmodium*. This malarial protist has a life cycle that includes the mosquito. Malaria causes recurrent episodes of fever and can be fatal.**

   a) Cut out the diagram and the boxes at the bottom of the page and arrange them in the correct place on a separate piece of paper to explain the stages of the malaria life-cycle diagram. Start with the box in bold. When you have checked your answers with your tutor, stick them into place.

   b) When you have done this, explain to your tutor why the fever from malaria comes and goes in recurrent episodes, and what a vector is.

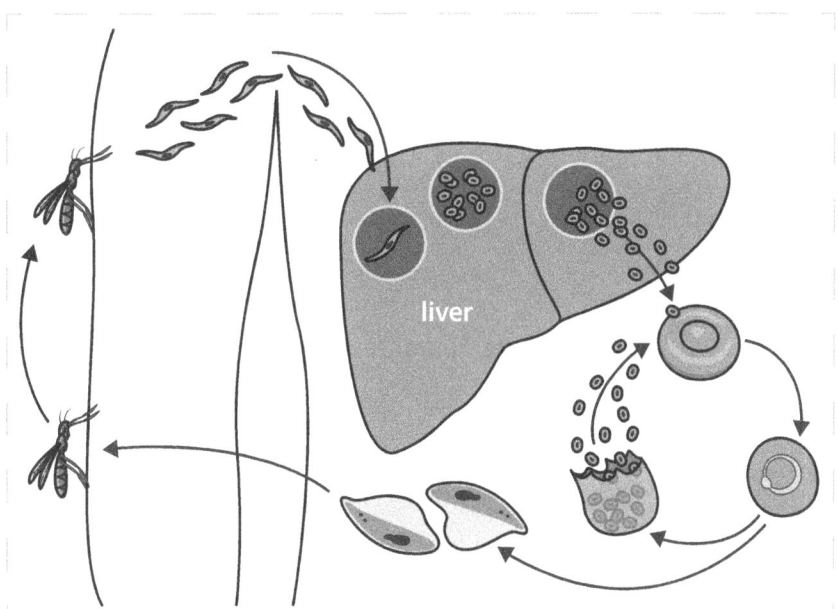

| **A malaria-carrying mosquito bites an uninfected person. The protist in the mosquito saliva infects the person.** | The protist then moves into the blood. It enters red blood cells. The protist multiplies by cell division inside the red blood cells. | The cycle of infection, division and release in the red blood cells is repeated every couple of days, causing recurring episodes of fever. | The malaria protist infects the salivary glands of the mosquito. |

| The malaria protist travels to the liver of the person, where the protist cells mature. | The red blood cells burst, releasing the malaria protist into the bloodstream. This triggers an immune response and a fever in the person. | When taking a blood meal, a mosquito may also take up some of the malaria protist. | The mosquito can now act as a vector and carry the malaria disease to infect another person. |

BIOLOGY HIGHER

Homework activity: Mosquitoes and malaria prevention

Time 15 mins

Learning objectives
- To explain how malaria can be controlled

Equipment
none

1. One way in which the spread of malaria may be controlled is by draining areas of standing water and covering up water storage tanks. Explain how this reduces the spread of malaria.

   _____

   _____

   _____

   _____

   _____

   _____

2. Name two other methods that can be used to control the spread of malaria and explain how each works.

   Method 1 _____

   _____

   _____

   Method 2 _____

   _____

   _____

12 Answers

Starter activity: Comparing pathogens

1.

Microorganism group	Characteristics of microorganism	Example of pathogen
virus	They are not cells; they are only made up of a protein coat and genetic material. They do not feed or respire, but rely on host cells for *reproduction*. They cause damage to host cells when *virus* particles burst out from cells.	*HIV, measles, tobacco mosaic virus (TMV)*
bacteria	*Prokaryotic* cells do not have a nucleus. They reproduce asexually by *binary* fission. Pathogenic bacteria produce *toxins* which damage body cells.	*salmonella*
fungus	Eukaryotic cells contain sub-structures such as a n*ucleus*. Single celled or multiple celled, include yeasts, m*oulds* and mushrooms May *reproduce* sexually or asexually using dispersal cells called spores	rose black spot
protist	E*ukaryotic* cells contain sub-structures such as a nucleus. Can be single celled or multiple celled; there are many different types. May reproduce sexually or asexually, often have complex life *cycles*	*Plasmodium*

Main activity: Rose black spot, a fungal disease

1. a) Growth is affected because photosynthesis is reduced. There are fewer leaves to photosynthesise and leaf tissue dies. Less sugar is made that can be used for plant growth.
   b) This kills the fungus so it cannot produce spores and infect more leaves.
   c) It is spread in the environment by water or wind. Spores can be blown in or brought in rain drops from elsewhere.
   d) Fungicide; fungicides are the usual chemical treatment for rose black spot disease.

Main activity: Malaria, a protist disease

1. a) The boxes are in the correct order, going down then across in the table.
   b) There are bouts of fever because the release of protist cells from the red blood cells occurs in a cycle. It is the free protists in the bloodstream that trigger the immune response of fever. A vector is something that carries a disease from one person to another.

Homework activity: Mosquitoes and malaria prevention

1. Student's own answer. Examples: mosquitoes lay their eggs in water/the larvae of mosquitoes live in water; so mosquitoes need access to standing water to breed/complete their life cycle; reduced breeding of mosquitoes will result in fewer mosquito adults; adult mosquitoes are vectors for malaria/spread the disease from person to person; if numbers of mosquitoes are reduced there will be a reduced spread of the disease
2. Student's own answer. Examples: use mosquito nets, which prevent mosquitoes from biting and infecting people while they sleep; use insecticides, especially around water bodies or in houses; they kill mosquitoes and reduce the chances of being bitten; use predators of larvae – stocking ponds with fish that eat mosquito larvae can reduce the numbers of vectors

13 Infection and response: Defence systems, vaccination, antibiotics and painkillers

Learning objectives

- To describe the defence systems of the human body against pathogens
- To explain the role of the immune system
- To explain how vaccination prevents illness in individuals and populations
- To explain the use of antibiotics and other medicines

Specification links

- 4.3.1.6
- 4.3.1.7
- 4.3.1.8

Starter activity

- **Human defence systems; 5 minutes; page 90**

  Ask the student to match the defence mechanism to the relevant part of the body. The second activity should provide an opportunity to check the student's understanding of specific versus non-specific immune systems.

Main activities

- **White blood cells; 15 minutes; page 91**

  Discuss the term 'antigen' with the student and make sure that the term is understood. The diagrams are then labelled and completed to show three ways in which white blood cells can fight pathogens. It is important to emphasise that antibodies and antitoxins are proteins that are specific to one type of antigen or toxin, so the correct one must be produced on exposure to the pathogen.

- **Vaccines, antibiotics and painkillers; 15 minutes; page 92**

  Explain the basic idea of the graph to the student and then allow them to complete the first task, adding letters to the graph. Discuss the answers and check understanding. The student then completes the second activity.

  Check their answers and knowledge. There are links between this content and lesson 31.

- **Herd immunity; 10 minutes**

  Provide the student with counters or coins of two different colours to represent vaccinated/unvaccinated people. Ask the student to use these to explain the importance of most people in a population being immunised. Students should explain how mass vaccination reduces the chances of spread of diseases because of the reduced proximity of susceptible people. Emphasise that if a critical proportion are immunised, even those who cannot be vaccinated get some protection.

Plenary activity

- **Note to tutor; 5 minutes**

  Ask the student to write down the most important thing that they learned in the lesson and one question that they still wish to ask. Discuss the question that the student asks.

Homework activity

- **Immunity and disease; 15 minutes; page 93**

  The student should complete the exam-style questions covering antibiotics and vaccination.

Support ideas

- **White blood cells** Show a short video clip of a phagocyte in action.
- **Vaccines, antibiotics and painkillers** Ask the student to name a disease which they have been, or could be, vaccinated against. Then ask them to explain why this wouldn't protect them from a different disease.

Extension ideas

- **White blood cells** Point out the differences in the nuclei of antibody-producing lymphocytes and phagocytes. Discuss the difference in specificity between the two types of white blood cell.
- **Vaccines, antibiotics and painkillers** Ask the student to use the graph to explain why you are unlikely to suffer from chicken pox twice.

Progress and observations

Starter activity: Human defence systems

Time 5 mins

Learning objectives
- To explain the non-specific defence systems of the human body
- To explain the role of the immune system

Equipment
none

1. Non-specific defence systems are those that act to prevent infection from or to kill any pathogen. These systems do not rely on recognition of the pathogen. Draw lines between the boxes to match the part of the body with the non-specific defence mechanism that it has against pathogens.

Part of the body	Defence mechanism
skin	Hairs lining the passages filter the air that is drawn in. They trap dirt and pathogens, stopping them from getting into the lungs.
eyes	They produce tears which contain a chemical that kills bacteria.
nose	They contain cells that produce a sticky mucus that traps pathogens. Air passages are also lined with cells that have cilia, which beat to sweep the mucus to the throat to be swallowed.
trachea and bronchi	This produces hydrochloric acid that kills bacteria in food.
stomach	The dead outer layer of cells acts as a physical barrier. Glands produce sebum, containing oils with antibacterial properties.

2. In fewer than 20 words, describe the role of the immune system.

_____
_____
_____

BIOLOGY HIGHER

Main activity: White blood cells

Time 15 mins

Learning objectives
- To explain the role of white blood cells in fighting disease

Equipment
- pencil

1. Discuss the meaning of the term 'antigen' with your tutor.

2. The diagrams below show three ways in which white blood cells help to defend the body against pathogens.

   a) Add the following labels in the appropriate places on the diagrams:

   lymphocyte (white blood cell)   antitoxin

   antibody   bacterial toxin

   bacterium   phagocyte (white blood cell)

   antigen

   b) Complete the diagrams to the right of the arrow to show how each of these processes defends against pathogens.

   producing antibodies

   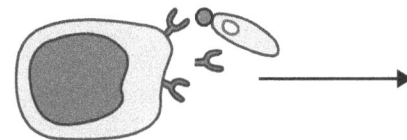

   producing antitoxins

   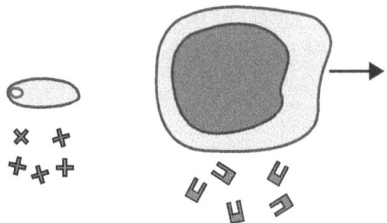

   phagocytosis

   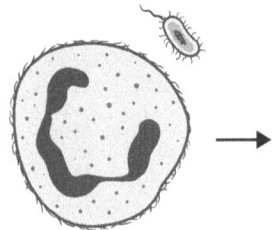

91

BIOLOGY HIGHER

AQA

Main activity: Vaccines, antibiotics and painkillers   Time 15 mins

Learning objectives
- To explain how vaccination prevents illness in individuals
- To explain the use of antibiotics and other medicines to treat disease

Equipment
none

1. The graph below shows the changes in antibody levels in a person's blood following vaccination and then exposure to the disease. Read the descriptions of events in the table, then write the letter for each event in the correct place on the graph.

A. Vaccination: a small amount of a dead or inactive form of a pathogen is introduced into the body, usually by injection.	B. The pathogen is destroyed quickly by the antibodies, so the person does not suffer from the disease. They are immune to it.	C. Memory cells that can produce the correct antibody respond quickly, multiplying and producing lots of antibodies
D. Exposure to the disease: the same pathogen, with the same antigen, enters the body.	E. Some white blood cells (memory cells) that can produce the antibody remain in the body.	F. This stimulates white blood cells to produce antibodies specific to the antigen.

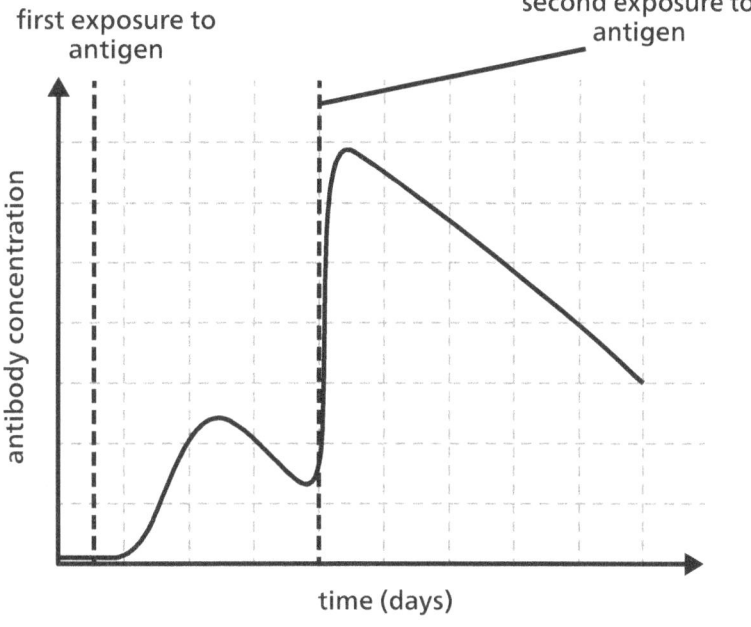

2. Complete the statements using a word from the word bank underneath.

Antibiotics are medicines that can kill pathogenic _____ .They cure bacterial diseases by killing infections inside the _____ . One example of an antibiotic is _____ . It is important that specific bacteria should be treated using antibiotics that are _____ .

Antibiotics became available as medicines in the middle of the 20th century, greatly reducing bacteria-related _____ . There are now concerns that deaths will rise again because of the emergence of strains that are antibiotic _____ . Antibiotics cannot kill pathogens that are _____ . Painkillers and other medicines do not kill pathogens, they are used to treat disease _____ .

It is difficult to develop drugs that kill viruses without also damaging the body's _____ . This is because viruses can only live and reproduce inside the body cell of their _____ .

Word bank: penicillin bacteria host specific deaths tissues body resistant viruses symptoms

BIOLOGY HIGHER

Homework activity: Immunity and disease

Time 15 mins

Learning objectives
- To explain how exposure to a disease and vaccination can provide immunity to a disease

Equipment
none

1. Sandra has been infected with flu. She goes to her doctor and asks for antibiotics. Her doctor tells her that she cannot prescribe antibiotics for flu because they will not work.

   Suggest why the antibiotics will not work to treat flu.

   _____
   _____

2. Older people may be given a vaccination against flu. They will need to have a flu vaccination every year. Most vaccines will protect against infection for many years.

   Explain why a new flu vaccine must be given every year.

   _____
   _____
   _____
   _____
   _____
   _____

3. Explain why it is important that a large proportion of a population is vaccinated against a disease such as measles.

   _____
   _____
   _____
   _____
   _____

13 Answers

Starter activity: Human defence systems

1. Skin: the dead outer layer of cells acts as a physical barrier; glands in the skin produce sebum, containing oils with antibacterial properties
   Eyes: produce tears which contain a chemical that kills bacteria
   Nose: hairs lining the passages filter the air that is drawn in; they trap dirt and pathogens, stopping them from getting into the lungs
   Trachea and bronchi: contain cells that produce a sticky mucus that traps pathogens; air passages are also lined with cells that have cilia, which beat to sweep the mucus to the throat to be swallowed
   Stomach: produces hydrochloric acid that kills bacteria in food
2. If a pathogen enters the body, the immune system will identify and then try to destroy the pathogen.

Main activity: White blood cells

1. An antigen is a chemical that is contained in a pathogen.
2. The diagram should show how antibodies bind to a specific antigen of specific shape. It might show that the antibody immobilises the bacteria (for phagocytes), neutralises it or causes cell lysis.
   Labels: lymphocyte, antibody, bacterium, antigen; producing antitoxins
   Diagram should show how antitoxins bind to toxins and neutralise them so that they no longer cause cell damage.
   Labels: lymphocyte, bacterium, bacterial toxin, antitoxin
   Diagram should show that the phagocytes engulf, ingest and destroy the bacteria using enzymes.
   Labels: phagocyte, bacterium, phagocytosis

Main activity: Vaccines, antibiotics and pain killers

1. A At first exposure, F first antigen peak, E decline of first peak, D second exposure, C second antigen peak, B decline of second peak
2. Antibiotics are medicines that can kill pathogenic *bacteria*. They cure bacterial diseases by killing infections inside the *body*. One example of an antibiotic is *penicillin*. It is important that specific bacteria should be treated using antibiotics that are *specific*. Antibiotics became available as medicines in the middle of the 20th century, greatly reducing bacteria-related *deaths*. There are now concerns that deaths will rise again because of the emergence of strains that are antibiotic *resistant*. Antibiotics cannot kill pathogens that are *viruses*. Painkillers and other medicines do not kill pathogens, they are used to treat disease symptoms. It is difficult to develop drugs that kill viruses without also damaging the body's *tissues*. This is because viruses can only live and reproduce inside the body cell of their *host*.

Homework activity: Immunity and disease

1. Antibiotics do not kill viruses/only kill bacterial infections
2. The DNA of the flu virus mutates/the antigens on its surface change; the antibodies produced in response to a vaccine will not match the new/mutated antigen; the old vaccine will not give immunity for a new strain of the virus; a new vaccine is needed each year containing antigens for the current strain of flu
3. Measles can have serious effects/be fatal; the larger the proportion of the population that is vaccinated, the less likely it is that the disease will spread; most people will be immune and unable to pass the disease on to someone else

14 Infection and response: Drug development and monoclonal antibodies

Learning objectives

- To describe how medicines were produced traditionally
- To identify some modern treatment approaches
- To describe the process of discovery and development of new medicines
- To describe how monoclonal antibodies are produced
- To describe some of the ways in which monoclonal antibodies can be used

Specification links

- 4.3.1.9
- 4.3.2.1
- 4.3.2.2

Starter activity

- **Traditional drugs; 5 minutes; page 96**

  Ask the student to complete parts one and two independently. Review these then ask the student to do part three verbally, defining both 'monoclonal' and 'antibody' separately, initially.

Main activities

- **Pre-clinical and clinical testing; 15 minutes; page 97**

  The student should complete the flow diagram independently. Then use the questions as a focus for discussion.

- **Monoclonal antibodies; 20 minutes; page 98**

  Give the student sufficient thinking time to read and take in the diagram, before asking them to add labels to show how monoclonal antibodies are made. If time allows, consider the ethical issues (see extension ideas). The third activity could be done verbally.

- **Monoclonal antibodies and cancer; 5 minutes**

  Ask the student to sketch a diagram to show how monoclonal antibodies target cancer cells but not healthy cells.

  A key point should be the complementary shapes of the antibody and binding site on cancer antigen, but different antigens on a healthy cell.

Plenary activity

- **Traffic lights; 5 minutes**

  Ask the student to look back through the sheets and to rate each section as either green (understand), amber (mostly understand) or red (not clear, need to revisit). Discuss any areas of uncertainty.

Homework activity

- **Using monoclonal antibodies; 15 minutes; page 99**

  Exam-style questions on monoclonal antibodies based around their use in pregnancy test kits. The student may need an explanation of the diagram.

Support ideas

- **Pre-clinical and clinical testing** Explain single-blind and double-blind tests using a pack of cards. Deal the student cards face down (blind), initially choosing which cards to give (single blind) then dealing them at random without looking.
- **Monoclonal antibodies** Use modelling material to model the idea of the complementary shape of an antibody and binding site, and that there can be more than one binding site on an antigen protein.

Extension ideas

- **Pre-clinical and clinical testing** Tell the story of the discovery of digitalis – how William Withering, after hearing of it as a herbal cure, tested it by trial and error on his patients to get the dose right. Ask the student to compare this with modern drug development.
- **Monoclonal antibodies** Discuss the ethics of producing monoclonal antibodies, for example the use of animals in research, production of human–mouse hybrid cells.

Progress and observations

Starter activity: Traditional drugs

Time 5 mins

Learning objectives
- To describe how medicines were produced traditionally
- To identify some modern treatment approaches

Equipment
none

1. In medicine, drugs are substances that are used to prevent or treat disease. Traditionally, drugs were extracted from plants and microorganisms. Complete the table to describe some traditional drugs and how they are produced.

Drug	Type of organism	Name of organism	Medical use	More information
Digitalis		foxglove	to treat _____ _____ disease	reduces irregular heartbeat and heart disease-related swelling of limbs
	plant	willow	_____ _____	extracted from the bark of the willow tree
Penicillin		penicillium mould	antibiotic – to treat bacterial diseases	discovered by _____ _____

2. Some drugs are still extracted from plants, but most new drugs are synthesised by chemists in which industry?
_____

3. Explain the term 'monoclonal antibody' to your tutor.

BIOLOGY HIGHER

Main activity: Pre-clinical and clinical testing

Time 15 mins

Learning objectives

- To describe the process of discovery and development of new medicines including pre-clinical and clinical testing

Equipment

none

1. New medical drugs must be tested and trialled before being used. They must be tested for toxicity (to see if they are safe), efficacy (are they effective in treating the disease) and dose (what is the smallest dose that will treat the disease). Complete the flow diagram by filling in the top two bullet points of each box, to describe the process of testing a new drug. Add detail next to the headings in bold.

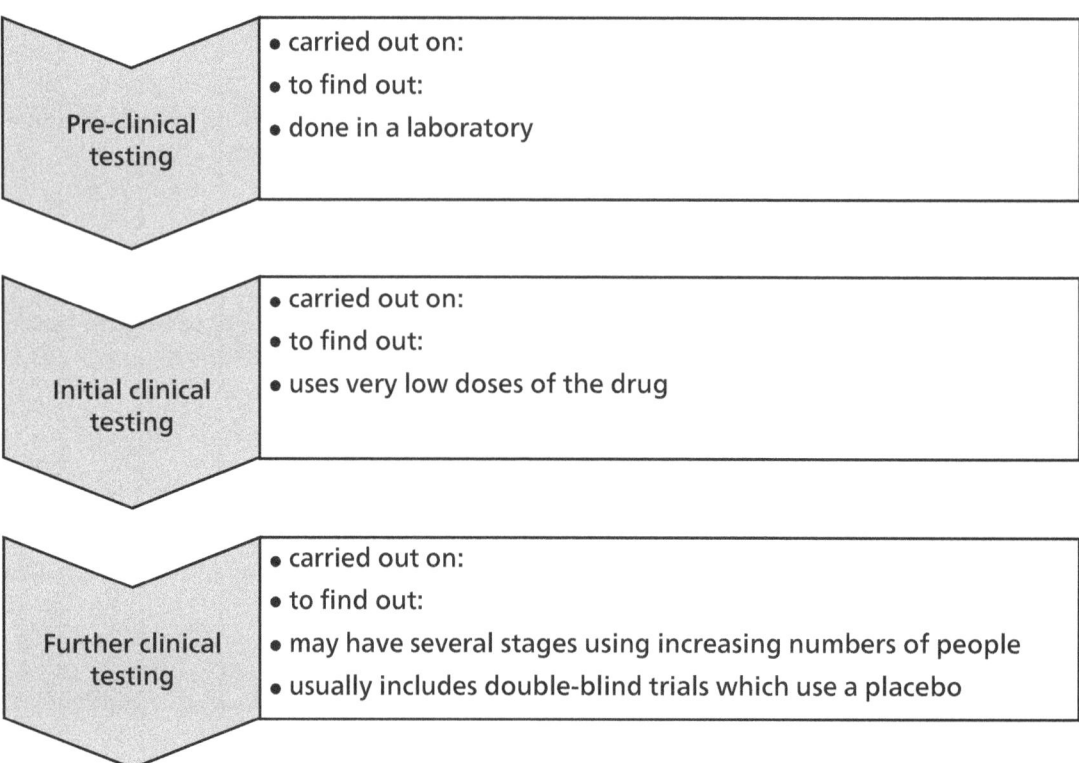

Pre-clinical testing
- carried out on:
- to find out:
- done in a laboratory

Initial clinical testing
- carried out on:
- to find out:
- uses very low doses of the drug

Further clinical testing
- carried out on:
- to find out:
- may have several stages using increasing numbers of people
- usually includes double-blind trials which use a placebo

2. Discuss the following questions with your tutor.

   a) What is meant by the terms placebo and double-blind trial?

   b) What is the main difference between pre-clinical and clinical testing?

   c) Do drugs always go through all these stages?

   d) Why are very low doses given at the start of clinical trials?

   e) What ethical issues may arise in the testing process?

   f) How would patients be selected for placebo or drug test groups?

   g) What would follow further clinical testing before the drug is licenced?

BIOLOGY HIGHER

Main activity: Monoclonal antibodies

Time 20 mins

Learning objectives
- To describe how monoclonal antibodies are produced
- To describe ways in which monoclonal antibodies can be used

Equipment
none

To produce monoclonal antibodies, a mouse is first injected with an **antigen**. This stimulates mouse **lymphocytes** to make an **antibody** that is specific to a binding site on that antigen. The lymphocytes are collected from the mouse, then **fused** together with a special kind of **tumour cells** to make a cell called a **hybridoma cell**. The hybridoma cell can both divide rapidly (like a tumour cell) and make the antibody (like the lymphocyte). A single hybridoma cell with the correct antibody is selected. This hybridoma is **cloned** to produce many identical cells that all produce the same antibody. **Lots of antibody** can then be **collected and purified**.

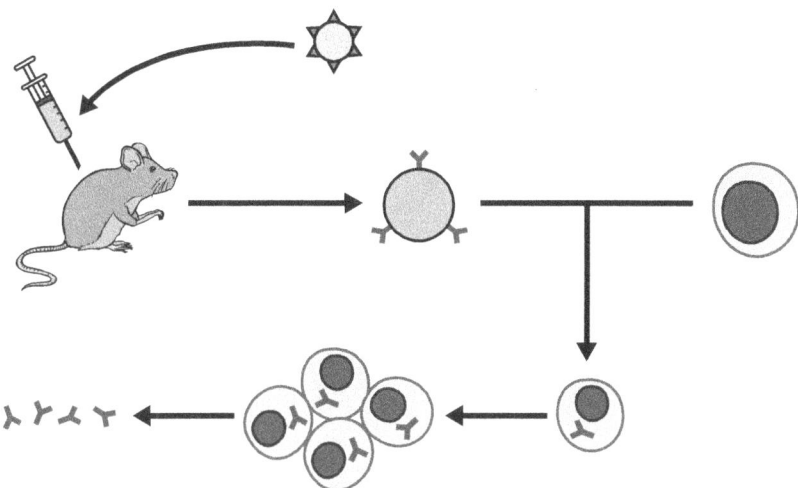

1. Use this description of how monoclonal antibodies are produced to add labels to the diagram above. Make sure that you include all the words in bold in your labels.

2. Complete the table to describe some examples of how monoclonal antibodies can be used.

What is the antigen?	Example of use	How do monoclonal antibodies work
pregnancy hormone (HCG)	diagnosis, for example in _____ tests	Monoclonal antibodies embedded in test kits bind to the specific antigen in a _____ sample, this causes a change in colour which allows measurement of the antigen levels in the blood.
hormone, pathogen or chemical	laboratory tests of _____	Monoclonal antibodies embedded in test kits bind to the specific antigen in a blood sample, this causes a change in colour which allows _____ of the antigen levels in the blood.
	in research to locate or identify specific molecules	Monoclonal antibodies are bound to a fluorescent dye and then introduced into the cell or tissue being tested. The antibodies bind to the specific antigen and the dye shows where that antigen is in the cell.
	to treat some diseases, such as cancer	The monoclonal antibody is bound to a chemical that will kill cancer cells or stop them growing and dividing (such as a _____ substance or a toxic drug). It delivers the substance to the cancer cells without harming other cells in the body.

3. Why are monoclonal antibodies not yet as widely used to treat disease as had been hoped when first developed?

_____

_____

BIOLOGY HIGHER

Homework activity: Using monoclonal antibodies

Time 15 mins

Learning objectives

- To explain the use of monoclonal antibodies

Equipment

none

The diagram shows one type of pregnancy test kit which contains two different monoclonal antibodies. A urine sample is placed on the sample area. It then moves along towards the test area. hCG hormone is only present in urine during pregnancy. If hCG is present, a dye in the test area will be activated and become coloured. The woman will know she is pregnant if she sees the colour in the test area.

The diagram shows what happens if the woman is pregnant.

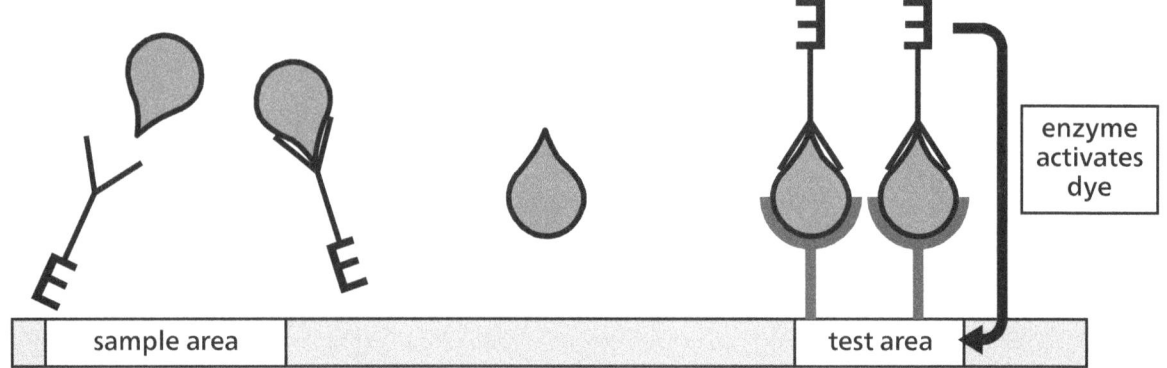

key

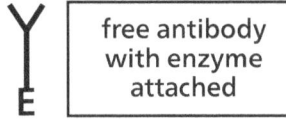 free antibody with enzyme attached

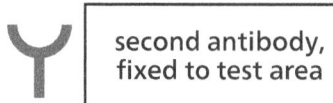 second antibody, fixed to test area

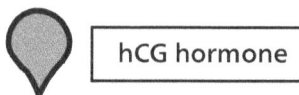 hCG hormone

1. Answer the following questions.

   a) What is a monoclonal antibody?

   _____
   _____

   b) Explain how the monoclonal antibodies bind to the hCG hormone.

   _____

   c) Explain how the two different monoclonal antibodies produce a positive test result when hCG is present.

   _____
   _____

14 Answers

Starter activity: Traditional drugs

1. Top row: plant, heart; middle row: aspirin, painkiller; bottom row: microorganism, Alexander Fleming
2. pharmaceutical
3. A monoclonal antibody is an immune protein produced by a single clone of cells, all genetically identical, produced by division of a single initial cell, all producing the same antibody. The antibody is specific to one binding site on one protein and will target a specific chemical or specific cells in the body.

Main activity: Pre-clinical and clinical testing

1. Pre-clinical testing: carried out on live animals or human cells and tissues grown in a lab, to find out drug toxicity/dose/efficacy
   Initial clinical testing: carried out on small numbers of healthy volunteers and on small numbers of patients, to find out drug toxicity/side effects
   Further clinical testing: carried out on large numbers of patients, to find out the optimum dose for the drug and its efficacy
2. a) See glossary
   b) Pre-clinical testing does not use people
   c) No, some will not be safe or effective enough to go to the next stage.
   d) In case drugs are very toxic
   e) Some people think that using animals in tests is wrong, but this may be balanced by a reduction in human suffering. It may not be ethical to continue to give a placebo if the drug is proving to be very effective.
   f) Random allocation to groups, so no one knows/there is no bias
   g) Peer review of results, to check claims of efficacy are true

Main activity: Monoclonal antibodies

1. Diagram label for each step in the process using words in bold
2.

What is the antigen?	Example of use	How do monoclonal antibodies work
Pregnancy hormone (HCG)	Diagnosis, for example in *pregnancy* tests	Monoclonal antibodies embedded in test kits bind to the specific antigen in a *urine* sample, this causes a change in colour which allows measurement of the antigen levels in the blood.
Hormone, pathogen or chemical	Laboratory tests of *blood*	Monoclonal antibodies embedded in test kits bind to the specific antigen in a blood sample, which causes a change in colour which allows *measurement* of the antigen levels in the blood.
The specific molecule to be tested for	In research to locate or identify specific molecules	Monoclonal antibodies are bound to a fluorescent dye and then introduced into the cell or tissue being tested. The antibodies bind to the specific antigen and the dye shows where that antigen is in the cell.
A molecule found on the surface of cancer cells but not healthy cells	To treat some diseases, such as cancer	The monoclonal antibody is bound to a chemical that will kill cancer cells or stop them growing and dividing (such as a *radioactive* substance or a toxic drug). It delivers the substance to the cancer cells without harming other cells in the body.

3. Monoclonal antibodies created more side effects than expected. They may trigger harmful immune reactions in the body.

Homework activity: Using monoclonal antibodies

1. a) Monoclonal antibodies are produced from a single clone of cells; they bind to one binding site on a specific antigen/protein.
   b) They have a complementary shape to the binding site on the antigen.
   c) The free antibody binds to a site on the hCG antigen, attaching an enzyme. The second antibody binds to a different binding site. It fixes the hCG onto the test area. The enzyme attached to the hCG can then activate the dye in the test area, causing the colour to appear.

BIOLOGY HIGHER

15 Infection and response: Plant diseases

Learning objectives

- To recall background knowledge about plant diseases
- To understand how plant diseases can be detected and identified
- To understand that plants can be harmed by ion deficiency conditions
- To describe physical and chemical plant responses

Specification links

- 4.3.3.1
- 4.3.3.2

Starter activity

- **Plant diseases – what I know; 5 minutes; page 102**

  Ask the student to record what they already know about plant diseases. They should be able to recall information about tobacco mosaic virus and rose black spot disease (lessons 11 and 12). This activity will be revisited in the plenary.

Main activities

- **Identifying plant diseases; 15 minutes; page 103**

  The student should label the plant diagram to record what signs may indicate disease. Add to the drawing to show leaf spots or deformations. Question 2 presents short questions based on required specification knowledge about the identification of plant diseases.

- **Plant deficiencies and defences; 20 minutes; page 104**

  Ask the student to complete the table then discuss the ion deficiency examples. Allow five minutes for the student to complete the crossword unassisted, then go through the answers to check their understanding. Finally, ask the student to highlight each clue in a different colour depending on whether the mechanism is physical (to prevent invasion), chemical or mechanical. Discuss the meaning of these terms. Then for question 4, ask the student to give you an example for each of the defences mentioned in the crossword.

- **Examples of defences; 5 minutes**

  Ask the student to think of an example for each of the defences mentioned in the crossword on the *Plant deficiencies and defences* activity sheet. Most will be straight forward. Mimicry examples include mimicry of the presence of insect eggs on leaves, dissuading insect species from laying their eggs there, or non-poisonous plants that mimic the appearance of unpalatable ones to trick herbivores into avoidance, such as Australian mistletoe.

Plenary activity

- **Plant diseases – what I know now; 5 minutes**

  Using a different colour pen, the student should add to their summary of knowledge on the starter sheet. This will show clearly the key areas of progression during the session.

Homework activity

- **Barberry defences; 15 minutes; page 105**

  The student should complete the extended exam-style question on the sheet. The main chemical defence of barberry is the alkaloid berberine.

Support ideas

- **Identifying plant diseases** Show photos of plants with disease symptoms to better aid memory.
- **Examples of defences** Show pictures of plants with key defences to aid memory. Barberry (spines and chemicals), eucalyptus (bark shedding) and mimosa (drooping when touched) are good examples.

Extension ideas

- **Identifying plant diseases** Discuss the life cycle of the aphid in more detail, especially the link between asexual reproduction and rapid colonisation of a plant host.
- **Plant deficiencies and defences** Ask the student to describe in more detail the link between lack of nitrates and stunted growth. Nitrogen is needed for proteins and DNA, essential for cell division and growth.

Progress and observations

Starter activity: Plant diseases – what I know

Time 5 mins

Learning objectives
- To recall background knowledge on plant diseases

Equipment
none

1. In the space below, record what you already know about plant diseases.

   You could use a bullet point list or a spider diagram. This could include:
   - causes of plant disease
   - examples of plant diseases
   - symptoms of plant diseases
   - how plants defend themselves.

Main activity: Identifying plant diseases **Time** 15 mins

Learning objectives
- To understand how plant diseases can be detected and identified

Equipment
- pencil

1. There are many signs that can help us to detect and identify diseases. Label or add on as many as you can think of on the plant below.

2. If a plant shows signs of a disease, how could you identify which disease it has? Write down three ways.
___
___
___

3. Aphids are insects that cause damage to many plants. Describe how aphids damage plants.
___
___
___
___

BIOLOGY HIGHER

Main activity: Plant deficiencies and defences

Time **20** mins

Learning objectives

- To understand that plants can be damaged by ion deficiency conditions
- To describe physical and chemical plant defence responses

Equipment

- three different coloured highlighter pens

1. Plants can be damaged by a range of ion deficiency conditions. Complete the table to describe the deficiency of two important ions needed by plants.

Ion	Symptom of deficiency	Reason for the symptom
Nitrate		lack of nitrogen needed for protein synthesis; used to make _____, the building blocks of proteins, essential for growth
		needed to make chlorophyll; it is chlorophyll that makes the leaf green

2. Plants may have a variety of defence responses both against disease and against being eaten by insects or grazing animals. Complete the crossword to describe some of these responses.

Across

1. this part of a tree may have layers of dead cells around it which fall off, stopping disease from entering the plant (4)
3. a harmless or tasty plant may use this method to trick animals into thinking they are a poisonous or bitter plant (7)
5. plants may produce these chemicals to deter herbivores (7)
7. cell walls are made of this strong material (9)
9. leaves may be covered in these, they may make it difficult to eat, or stop insects getting close to the leaf surface (5)

Down

2. this type of chemical may be made by plants to kill bacteria (13)
4. this outer surface layer of leaves may have a tough waxy coating (7)
6. these modified stems are sharp and discourage animals from eating the plant (6)
8. these may droop or curl when touched in some species of plant, which may discourage animals from eating them (6)

3. Plant defence responses can be grouped into physical defence responses to resist invasion of microorganisms, chemical defences or mechanical adaptations against animals. Choose a colour to represent each of these, then highlight each crossword clue above to show which type of response it represents.

4. Describe to your tutor an example for each of the defences mentioned in the crossword.

104

BIOLOGY HIGHER

Homework activity: Barberry defences

Time: 15 mins

Learning objectives
- To describe physical and chemical plant defence responses

Equipment
none

The barberry plant is a popular hedge plant in some countries. Here is a description of barberry taken from a garden plant catalogue.

> "There are many varieties of barberry plant. They have attractive, colourful leaves. These are not damaged in windy conditions and keep their waxy shine. They make an excellent security hedge because the stems are covered in sharp spines. Barberry was once widely used as a medicinal plant, extracts of the plant were used to treat skin infections. Barberry is resistant to damage from disease or animals, and is a good choice for your garden hedge."

1. Use the information provided here, along with your own knowledge, to explain why the barberry plant is resistant to damage from disease and animals.

[6 marks]

 **BIOLOGY** HIGHER

15 Answers

Starter activity: Plant diseases – what I know

1. Student's own answers. Ensure they have included information about tobacco mosaic virus and rose black spot fungus.

Main activity: Identifying plant diseases

1. Labels should include: stunted growth, spots on leaves, areas of decay (rot), growths, malformed stems or leaves, discolouration and the presence of pests
2. Refer to a gardening manual or website; take infected plants to a laboratory to identify the pathogen; use testing kits that contain monoclonal antibodies which are specific to plant disease antigens
3. They have long slender mouthparts that they use to pierce stems and leaves, reaching the phloem and sucking out sugar solution. They reproduce rapidly and may deprive the plant of energy that was made by photosynthesis, hence stunting growth. They exude a sticky solution that coats leaves and encourages fungal growth/infections. They may carry viral plant diseases. Some aphid species inject poisons and may cause deformed growth/galls.

Main activity: Plant deficiencies and defences

1. Top row: stunted growth, amino acids; bottom row: magnesium, chlorosis
2. & 3. P = physical defence, C = chemical defence, M = mechanical adaptation
   Across: 1. bark (P), 3. mimicry (M), 5. poisons (C), 7. cellulose (P), 9. hairs (M)
   Down: 2. antibacterial (C), 4. cuticle (P), 6. thorns (M), 8. leaves (M)
4. Student's own answers

Homework activity: Barberry defences

The following table provides guidance on what a Level 3, 2 or 1 answer to this question would look like and the number of marks each would attract.

L3	A clear, logical and coherent answer, with no significant irrelevant information. The student understands which features protect against disease or animals and links features accurately to explanations of how disease or damage is prevented.	5–6 marks
L2	A partial answer with some errors, irrelevant features, some ineffective explanation or linkage.	3–4 marks
L1	One or two relevant features identified but little in the way of logical explanation.	1–2 marks
	Indicative content • Leaves not damaged by wind – are tough • Cellulose cell walls provide physical barrier • Cellulose may be thickened in response to infection • Leaves have a waxy cuticle • Waxy cuticle is a tough physical defence • Water runs off waxy cuticle so fungal spores/bacteria are less likely to grow/survive • Spines on stems are a mechanical defence • Spines deter herbivores/insects from eating the plant because they are sharp/tough • Medicinal extract treated skin disease • So extract may be an antibacterial/antimicrobial chemical • This is a chemical plant defence response • Chemical may taste bad/be a poison to deter herbivores	

 **BIOLOGY HIGHER**

16 Bioenergetics: Photosynthesis

Learning objectives

- To describe the reaction of photosynthesis
- To explain the effects of limiting factors on the rate of photosynthesis
- To explain how the rate of photosynthesis can be measured in aquatic plants
- To interpret graphs to find the rate of photosynthesis
- To describe the uses of glucose from photosynthesis

Specification links

- 4.4.1.1
- 4.4.1.2
- 4.4.1.3
- MS 3d, 4a, 4d

Starter activity

- **Photosynthesis reactions; 5 minutes; page 108**
  This activity provides a reminder of the basic reaction and process of photosynthesis. Allow the student to complete this independently then discuss their answers.

Main activities

- **Limiting factors in photosynthesis; 15 minutes; page 109**
  In this activity, the student sketches and interprets graphs to show the effect of limiting factors and combinations of limiting factors on the rate of photosynthesis. Additional paper will be needed. The explanations in questions 1–4 could be done verbally or as annotation to the graph. It is important to mention chlorophyll as a limiting factor.

- **Photosynthesis investigations; 15 minutes; page 110**
  This activity is based around required practical activity 6 (investigate the effect of light intensity on the rate of photosynthesis using an aquatic organism). Ask the student to answer question 1 verbally. For question 2, emphasise that distance from the lamp does not give a linear decrease in light intensity. It is better to use a light meter to set light levels, or calculate and plot $1/d^2$. For question 3, point out that one to seven minutes is the maximum rate, which is why this range was used. There was a lag at the beginning before oxygen started to be released and the rate was slowing at the end.

- **Products of photosynthesis; 10 minutes**
  Ask the student to name or jot down as many uses of glucose from photosynthesis as they can. This is reviewed in the homework activity and the required knowledge is covered in the answers to the homework.

Plenary activity

- **On a scale of 1 to 10; 5 minutes**
  Ask the student to score their understanding of the topic of photosynthesis on a scale of 1 to 10. Ask them to suggest what they need to work on to get to the next level. If they give a score of 10, get them to ask you some tricky questions.

Homework activity

- **Uses of glucose in plants; 20 minutes; page 111**
  Ask the student to complete the spider diagram.

Support ideas

- **Limiting factors in photosynthesis** If the student is struggling to sketch the temperature graph, remind them that photosynthesis includes enzyme-driven reactions.
- **Photosynthesis investigations** If the student has not carried out this investigation, there are numerous video clips available online that they could view. A search for 'pondweed photosynthesis video' will give suitable results.

Extension ideas

- **Limiting factors in photosynthesis** Ask the student to suggest what is likely to be a limiting factor on a winter's day in the UK (temperature/light) and on a bright sunny day in a tropical forest (CO_2).
- **Photosynthesis investigations** Ask the student to sketch or describe how to collect oxygen over water and measure the volume using a gas syringe or graduated capillary tubing.

Progress and observations

Starter activity: Photosynthesis reactions

Time 5 mins

Learning objectives
- To describe the reaction of photosynthesis

Equipment
none

1. Decide if the following statements about photosynthesis are true or false. Circle T or F next to each one.

    a) Photosynthesis allows plants to make their own water by harnessing the Sun's energy.   T F

    b) Photosynthesis is an exothermic reaction.   T F

    c) Light transfers energy to the chloroplasts during photosynthesis.   T F

    d) Our atmosphere is only rich in oxygen due to photosynthesis happening over millions of years.   T F

2. Complete the equation that represents photosynthesis.

$$\text{carbon dioxide} + \underline{} \xrightarrow{\text{light}} \text{glucose} + \underline{}$$

3. Write the chemical symbols for:

    a) carbon dioxide _____

    b) glucose _____

BIOLOGY HIGHER

Main activity: Limiting factors in photosynthesis   **Time** 15 mins

Learning objectives
- To explain the effects of limiting factors on the rate of photosynthesis

Equipment
- extra paper, ideally graph paper
- coloured pencils

The rate of photosynthesis is influenced by four main factors. These are temperature, light intensity, carbon dioxide concentration and the amount of chlorophyll. These factors interact and any one of them may be the factor that limits photosynthesis. A graph showing how light intensity affects the rate of photosynthesis (when other factors are kept constant) is shown below.

1 **Explain to your tutor why:**

   a) The graph is linear initially.

   b) The graph levels off and eventually shows no further increase with increasing light intensity.

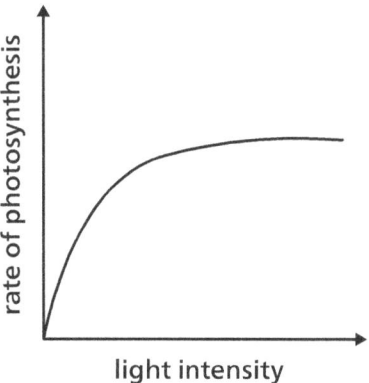

2. On a separate piece of paper, sketch a graph to show the effect of temperature on the rate of photosynthesis (if all other factors are kept the same). Explain the shape of your graph to your tutor.

3. Sketch a separate graph to show the relationship between carbon dioxide concentration and the rate of photosynthesis. Explain the shape of your graph to your tutor.

4. On the graph from question 1, add an additional line to show the effect of repeating the experiment at an increased temperature and a higher carbon dioxide concentration.

5. A company producing tomatoes grown in greenhouses wants to maximise their production and maintain profits. They use a lighting system.

   a) Suggest other ways that the producer might increase photosynthesis rates to improve production.

   _____
   _____

   b) The producer might decide not to use the conditions of CO_2, temperature and light that give the absolute maximum rate of photosynthesis. Why might they decide this?

   _____
   _____

BIOLOGY HIGHER

Main activity: Photosynthesis investigations Time 15 mins

Learning objectives
- To explain how the rate of photosynthesis can be measured in aquatic plants
- To interpret graphs to find the rate of photosynthesis

Equipment
- calculator

One way of measuring the rate of a reaction is to measure the rate at which products are made. Oxygen is a product of photosynthesis. In aquatic plants, oxygen bubbles can be seen and collected as they are released.

1. Look at the diagram showing one such investigation. The student recorded the rate at which bubbles were released from pondweed. The investigation was carried out in a darkened room.

   a) Why was the sodium bicarbonate added to the water?

   b) What was the purpose of the tank of water between the lamp and the pondweed?

   c) What other piece of equipment would be required if the rate of bubble release is to be measured?

   d) Name two important factors that should be controlled in this investigation.

   e) Describe a more precise way in which the amount of oxygen produced could be measured.

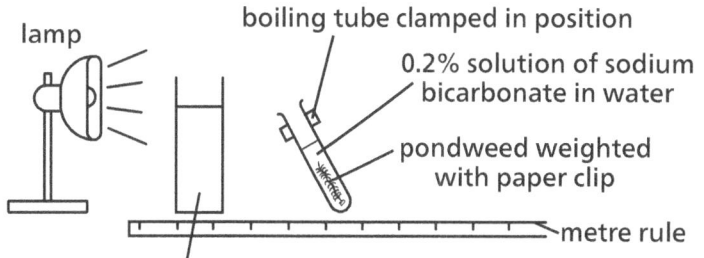

2. The student varied the light by recording the rate of oxygen production at increasing distances from the lamp. Light intensity changes with distance from the lamp following the inverse square law so that for a particular light source:

   Relative light intensity (arbitrary units) = $1/d^2$

   a) Calculate relative light intensity for the distances shown in the table. The first one has been done for you.

   b) Look at the values for 10 cm and 20 cm from the lamp. Describe the change in light intensity as distance doubles.

Distance from lamp (cm)	5	10	15	20
Relative light intensity	0.04			

   _____

   _____

   _____

   _____

3. A similar experiment was carried out where the volume of oxygen was measured over time.

   The graph shows the results at one level of light intensity. Use the graph to calculate the rate of photosynthesis between one minute and seven minutes.

   _____

   _____

   _____

   _____

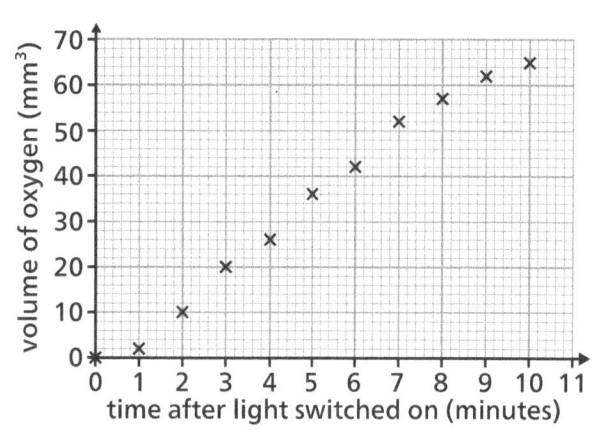

BIOLOGY HIGHER

Homework activity: Uses of glucose in plants

Time 20 mins

Learning objectives
- To describe the uses of glucose from photosynthesis

Equipment
none

1. Draw a spider diagram to show how the glucose produced in photosynthesis may be used by a plant. Use the framework provided as a starting point. Add a further point with detail or explanation for each use.

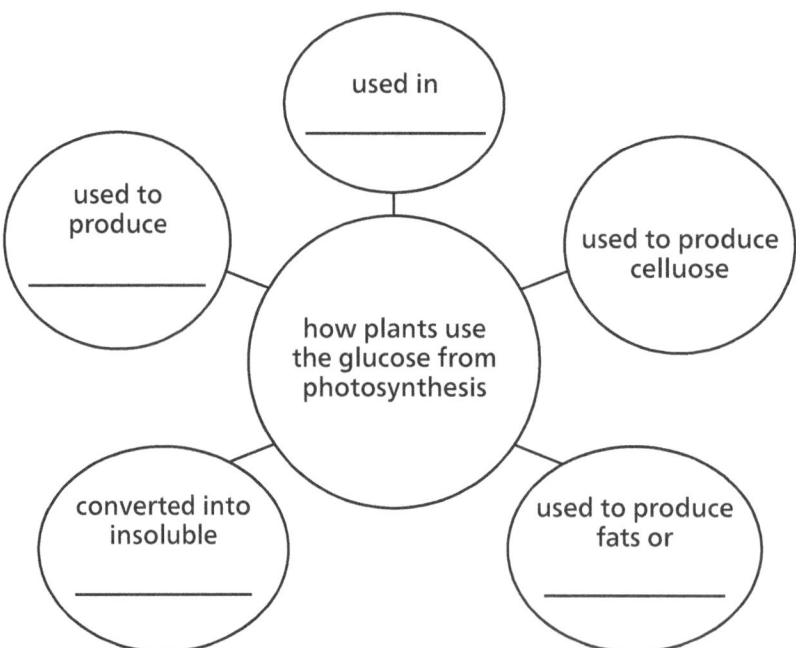

16 Answers

Starter activity: Photosynthesis reactions

1. a) F (uses water)
   b) F (is endothermic)
   c) T
   d) T
2. Water and oxygen
3. a) CO_2
   b) $C_6H_{12}O_6$

Main activity: Limiting factors in photosynthesis

1. a) The rate of photosynthesis is proportional to the light intensity.
   b) Another factor is limiting photosynthesis.
2. Student's own answer. Check they have sketched a graph typical of change in rate of an enzyme-driven reaction; rate increasing with temperature up to around 40 °C then declining to zero
   Explanation: increased reaction rate due to increased kinetic energy/particle collisions up to optimum, denaturing of enzymes during photosynthesis at higher temperatures causes decline in rate
3. Student's own answer. Check they have sketched a graph similar in shape to the light graph.
   Explanation: CO_2 is a raw material of photosynthesis, so increased rate of photosynthesis is proportional to increase in CO_2 concentration, at the plateau another factor is limiting photosynthesis
4. Line lies above the light-only line, with initial linear increase in rate, then plateaus
5. a) Aim for the optimum temperature and increase CO_2 levels. This can be done by using paraffin heaters that burn oil and release CO_2. The temperature in summer may need to be reduced by ventilation if it goes above the optimum. Provide sufficient mineral ions especially Mg to maximise chlorophyll content. Treat infections that might reduce chlorophyll levels or block light from the leaf.
   b) The costs of electricity/fuel might be too great, and the producer would aim to maximise profit, not simply photosynthesis rate. Beyond a certain point an increase in light, temperature and CO_2 may not be proportional to increased photosynthesis and would give diminished returns, as the amount of available chlorophyll would start to limit the reaction.

Main activity: Photosynthesis investigations

1. a) To provide a source of carbon dioxide for the pondweed
   b) To absorb any heat from the lamp
   c) Stop clock
   d) Temperature and size of the piece of pondweed (mass/number of leaves/length)
   e) Collect gas and measure the volume using gas syringe or capillary tubing
2. a) 0.01; 0.044; 0.0025
   b) As the distance doubles, the relative light intensity falls to a quarter.
3. Student should draw a line of best fit between one and seven minutes; rate should be 7.1 to 7.2 mm³ min⁻¹ depending on line of best fit

Homework activity: Uses of glucose in plants

1. Uses of glucose should include: used in respiration (to release energy for metabolic reactions), converted into insoluble starch (for storage, starch is a polymer of glucose), used to produce fat or oil (energy rich molecules, for storage), used to produce cellulose (strengthens the cell wall, polymer of glucose), used to produce amino acids (protein synthesis, also requires nitrate ions absorbed from soil)

17 Bioenergetics: Respiration

Learning objectives

- To describe and compare the processes of aerobic and anaerobic respiration
- To explain the response of the body to exercise
- To explain how oxygen debt develops and the recovery process
- To explain the importance of simple molecules in metabolism
- To understand the importance of respiration in metabolism

Specification links

- 4.4.2.1
- 4.4.2.2
- 4.4.2.3

Starter activity

- **Cellular respiration; 5 minutes; page 114**

  Ask the student to complete questions 1–3 independently. Question 4 should be answered verbally.

Main activities

- **Aerobic and anaerobic respiration; 15 minutes; page 115**

  In this activity, the student completes a number of short questions covering the basic facts about anaerobic and aerobic respiration. These should be completed one at a time, allowing time for discussion if their knowledge is not secure.

- **Exercise and respiration; 15 minutes; page 116**

  The student should complete question 1 verbally. Then discuss the principles of oxygen debt. Ask them to answer questions 2–5 then check their understanding. Question 6 should be done verbally. Encourage the student to consider independent variables, dependent variables and controlled variables.

- **Metabolic reactions; 10 minutes**

  Introduce metabolism as the sum of all the reactions in a cell or the body. Respiration is a metabolic reaction and supplies energy for other reactions. Ask the student to recall the following metabolic reactions and use simple diagrams to show how molecules are built up or broken down: the conversion of glucose to starch, glycogen and cellulose; the formation of lipid molecules from glycerol and fatty acids; the use of glucose and nitrate ions to form amino acids which are used to synthesise proteins; breakdown of excess proteins to form urea for excretion. This knowledge is covered elsewhere in the specification but is brought together here. It is revisited in the homework activity.

Plenary activity

- **Events during exercise; 5 minutes**

  The student should construct a flow diagram to show the changes that take place in the body during a 100 m sprint. Include changes in breathing, heart rate and concentrations of the reactants and products of respiration in the muscles.

Homework activity

- **Metabolism; 30 minutes; page 117**

  The student should complete the questions and table to summarise their knowledge about metabolism.

Support ideas

- **Aerobic and anaerobic respiration** It may help the student's understanding of the transfer of energy and the exothermic nature of respiration to compare respiration to burning. Ask the student to suggest similarities and differences.
- **Exercise and respiration** Ask the student to describe how they feel after exercise. If they include 'being hot and sweaty' link this back to the exothermic nature of respiration. Ask them to breathe as they would after exercise and describe the difference with resting. Ask them to take their pulse and describe how it would change.

Extension ideas

- **Aerobic and anaerobic respiration** Ask the student to sketch a mitochondrion and remind them that the folded membrane provides a greater surface area for respiration.
- **Exercise and respiration** Extend the student's understanding of the nature of energy transfer as the production of ATP.

Progress and observations

Starter activity: Cellular respiration

Time 5 mins

Learning objectives
- To describe the process of cellular respiration

Equipment
none

1. Fill in the gaps to complete the following sentences about respiration. Use the word bank to help you. Words may be used more than once.

   Respiration takes place continuously inside all living cells. The function of respiration is to _____ energy from food. Respiration provides the _____ to perform all living processes. Respiration in cells can take place _____ (using oxygen) or _____ (without oxygen). Aerobic respiration uses _____ to oxidise food and transfers _____ amounts of energy. Anaerobic respiration does not require _____ but transfers _____ energy than aerobic respiration. Anaerobic respiration takes place during vigorous _____ because the body is unable to supply the cells with enough oxygen. This process supplies energy but also causes the build-up of _____ in muscles. Lactic acid build up causes _____ fatigue.

   Word bank: muscle   transfer   less   large   aerobically   anaerobically   oxygen   exercise   energy   lactic acid

2. What substances are needed for aerobic respiration?
   _____

3. What substances are produced by aerobic respiration?
   _____

4. Explain how the substances listed in questions 2 and 3 are supplied to cells or removed from cells.
   _____
   _____
   _____
   _____

Main activity: Aerobic and anaerobic respiration

Time: 15 mins

Learning objectives
- To describe the process and function of aerobic respiration
- To compare anaerobic respiration in animal, plant and yeast cells
- To compare the processes of aerobic and anaerobic respiration

Equipment
none

1. Cellular respiration is an exothermic reaction. What does this mean?

   _____

2. Respiration releases energy from food. List three things that organisms might use this energy for.

   _____

3. Write the word equation for aerobic respiration.

   _____

4. Rewrite the equation for aerobic respiration using chemical symbols.

   _____

5. Write a word equation to represent the anaerobic respiration that takes place in muscles.

   _____

6. Why does anaerobic respiration release less energy from every molecule of glucose than aerobic respiration?

   _____

7. Anaerobic respiration in plant and yeast cells is different than in animal cells. Write a word equation to represent anaerobic respiration in plant and yeast cells.

   _____

8. Anaerobic respiration in yeast cells is called fermentation. Explain why each of the products of fermentation, carbon dioxide and ethanol, are important in the food and drinks industry.

   _____

   _____

9. Complete the table to summarise ideas about anaerobic and aerobic respiration.

Process	aerobic respiration	anaerobic respiration (animal)	fermentation (yeast)
Reactants			
Products			
Amount of energy released			

BIOLOGY HIGHER

Main activity: Exercise and respiration Time 15 mins

Learning objectives
- To describe the response of the body to exercise
- To explain how oxygen debt develops and the recovery process

Equipment
- coloured pencils for shading

1. During exercise your body has an increased demand for energy in order to move your muscles. This means that the muscles must be supplied with a greater amount of oxygen, allowing aerobic respiration to increase. Describe to your tutor three changes that take place in your body to provide more oxygenated blood to muscles when you start to exercise.

2. When we exercise it takes time for the body to respond to changes in oxygen and carbon dioxide levels. If insufficient oxygen is supplied, anaerobic respiration takes place in muscles. Build-up of lactic acid occurs, which creates an oxygen debt.

Add the following labels to the graph:
- start of exercise
- end of exercise
- resting oxygen uptake

Then shade and label areas of the graph to show:
- oxygen deficit during exercise
- oxygen debt

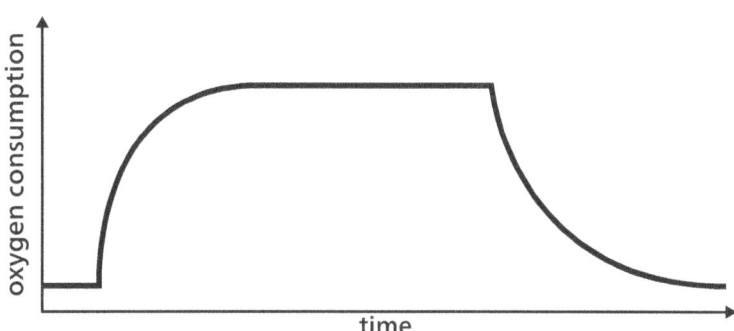

3. After exercise, our breathing and heart rate will remain high for a while to supply more oxygen because of the oxygen debt. Write a sentence to explain what this 'oxygen debt' is.

_____

4. What happens to the lactic acid that has built up in the muscle cells?

_____

5. A runner tries to sprint for 400 m. He starts fast and his heart rate and breathing rate increase, but he soon begins to slow down. When he stops, his legs feel wobbly and weak. Why does this happen?

_____

6. You are provided with a stop clock and a treadmill. Discuss with your tutor how you might use this equipment to investigate the effect of intensity of exercise on heart rate.

BIOLOGY HIGHER

Homework activity: Metabolism

Time 30 mins

Learning objectives
- To describe some metabolic reactions
- To understand the importance of respiration in metabolism

Equipment
- pencil
- spare paper

1. Metabolism is the sum of all the reactions in a cell or the body. What type of biological molecule acts as a catalyst for reactions within the body?

   _____

2. Many reactions in the body require energy. What process transfers energy to allow these reactions to take place?

   _____

3. The table lists some important metabolic reactions. Fill in the gaps to complete the table. One row has been completed for you.

Metabolic reaction	Reactants	Products	Synthesis or breakdown reaction?
conversion of glucose to starch, glycogen and cellulose	glucose	starch, glycogen or cellulose	synthesis
the formation of lipid molecules from glycerol and fatty acids	one molecule of _____ , three molecules of _____		
the use of glucose and nitrate ions to form amino acids			
the conversion of amino acids into proteins			
aerobic respiration			
the conversion of excess proteins into urea for excretion			

4. Starch, glycogen and cellulose are polymers of glucose; proteins are polymers of amino acids. On a separate piece of paper, draw a simple diagram to represent:

   a) A starch molecule

   b) A protein molecule

5. Lipid molecules are not polymers. They are not made from many similar subunits joined together. On a separate piece of paper, draw a simple diagram to represent a lipid molecule.

BIOLOGY HIGHER

17 Answers

Starter activity: Cellular respiration

1. Transfer, energy, aerobically, anaerobically, oxygen, large, oxygen, less, exercise, lactic acid, muscle
2. Oxygen and glucose
3. Carbon dioxide and water
4. They are supplied and removed via blood; glucose from digestion of food, oxygen from lungs carried in red blood cells, carbon dioxide to lungs; ideas of exchange of gases into/out of cells by diffusion

Main activity: Aerobic and anaerobic respiration

1. Exothermic reactions transfer heat energy to the surroundings. Respiration releases heat.
2. Chemical reactions to build larger molecules, movement, keeping warm
3. Glucose + oxygen → carbon dioxide + water
4. $C_6H_{12}O_6 + 6O_2 \rightarrow 6CO_2 + 6H_2O$
5. Glucose → lactic acid
6. In anaerobic respiration, the oxidation of glucose is incomplete, so there is still a lot of energy remaining in the products.
7. Glucose → ethanol + carbon dioxide
8. CO_2 is important in bread making; it causes bread to rise giving it a softer texture. Fermentation to produce ethanol is used to make alcoholic drinks.
9. Aerobic respiration: reactants: glucose and oxygen; products: carbon dioxide and water; energy: large amounts of energy
   Anaerobic (animal): reactants: glucose; products: lactic acid; energy: small amounts of energy
   Fermentation: reactants: glucose; products: carbon dioxide and ethanol; energy: small amounts of energy

Main activity: Exercise and respiration

1. Increased heart rate, increased breathing rate and increased breath volume
2. Labels in correct place, initial 'triangle' above graph shaded and labelled as oxygen deficit, section below graph after exercise shaded and labelled as oxygen debt
3. Oxygen debt is the amount of extra oxygen the body needs after exercise to react with the accumulated lactic acid and remove it from cells.
4. Blood flowing through the muscles transports the lactic acid to the liver where it is converted back into glucose.
5. During long periods of vigorous activity oxygen supply does not meet demand. The lactic acid build up causes muscles to become fatigued and stop contracting efficiently.
6. Dependent variable: measure heart rate, for example using electronic means/chest strap or count pulse for 15 seconds immediately after exercise stops; calculate beats per minute; independent variable: test at least five different speeds of running or gradient of slope; control variables: time for exercise, rest time, factors linked to subjects such as gender, age, fitness; repeat at least twice/three subjects at each level of the independent variable; consider safety risks

Homework activity: Metabolism

1. enzymes
2. respiration
3.

the formation of lipid molecules from glycerol and fatty acids	one molecule of glycerol, three molecules of fatty acids	lipid	synthesis
the use of glucose and nitrate ions to form amino acids	glucose and nitrate ions	amino acids	synthesis
the conversion of amino acids into proteins	amino acids	proteins	synthesis
aerobic respiration	glucose and oxygen	carbon dioxide and water	breakdown
the conversion of excess proteins into urea for excretion	proteins	urea	breakdown

4. a) Chain of identical molecules representing glucose; b) A chain of molecules that are not identical, representing different amino acids
5. A simple diagram showing the shape of a lipid as three long fatty acid molecules attached to one glycerol molecule

118

18 Homeostasis and response: Homeostasis and the nervous system

Learning objectives

- To explain the principles of homeostasis
- To describe body control systems (receptor, coordination centre, effector)
- To explain how thermoregulatory mechanisms lower or raise body temperature
- To explain how the structures in a reflex arc are adapted to their function
- To explain how the structure of the nervous system is adapted to its functions

Specification links

- 4.5.1
- 4.5.2.1
- 4.5.2.4

Starter activity

- **Principles of homeostasis; 5 minutes; page 120**

  Allow the student to complete the sheet independently. Questions 1 and 2 should not present any problems for higher students. Give prompts for question 3 if names cannot be recalled. Make sure that the student mentions optimum conditions for enzymes in question 1.

Main activities

- **The human nervous system; 15 minutes; page 121**

  The student should complete the first question independently. Check their understanding, then ask them to complete the drawing in question 2; prompts may be needed here. Allow the student to watch their own iris constriction/dilation response in a mirror – use a normal well-lit room only, do not shine a light in their eye. Discuss the reflex.

- **Thermoregulation; 15 minutes; page 122**

  Ask the student to complete question 1 independently on the sheet, then check their understanding. The student should then give verbal answers to the remaining questions.

- **Measuring reaction time; 10 minutes**

  Give the student a long ruler (a metre ruler if possible) and ask them to describe how they would use it investigate the effect of a given factor (such as caffeine or age) on human reaction time – a required practical. Ask them to describe how variables would be measured, changed or controlled. Discuss the potential sources of error in the experiment (can the subject tell when it will be dropped, not recording time directly). Discuss other methods of measuring reaction times, for example electronic methods based on visual signals.

Plenary activity

- **Guess the word; 5 minutes**

  The student should choose keywords from the lesson and describe them without using the word. Suggest which word is being described. Use this to assess the student's understanding of key terms.

Homework activity

- **The reflex arc; 15 minutes; page 123**

  The homework is a table completion activity and multiple choice question on the reflex arc.

Support ideas

- **The human nervous system** Show the student diagrams of the three types of neurone to support their understanding of the differences between them.
- **Thermoregulation** Show a short video clip of vasodilation to avoid the misconception that capillaries move in the skin.

Extension ideas

- **The human nervous system** Ask the student to draw the position of the cell bodies of each neurone onto the diagram.
- **Measuring reaction time** Show online reaction time measuring devices and discuss their advantages over the ruler method. Search for 'reaction time test' to find several examples.

Progress and observations

 # BIOLOGY HIGHER

Starter activity: Principles of homeostasis   Time 5 mins

Learning objectives
- To explain the principles of homeostasis
- To describe body control systems (receptor, coordination centre, effector)

Equipment
none

Homeostasis is the regulation of the internal conditions of a cell or organism. Examples of conditions that are regulated include blood glucose concentration, body temperature and water levels.

1. In homeostasis, internal conditions are kept constant, or within narrow boundaries. Why must internal conditions be maintained in this way?

    _____

    _____

2. Homeostasis involves automatic control systems. These can be nervous responses or chemical responses. The boxes in the centre compare nervous or chemical responses. Draw an arrow from each central box to the type of response that it describes.

| nervous responses | | chemical responses |

- longer lasting response
- faster response
- involves electrical impulses
- involves hormones released from glands
- information sent in the blood plasma
- information sent along cells

3. Both nervous control systems and chemical control systems are made up of three components or parts. Complete the table below to show the names of these three parts.

Part of control system	Type of structure	Function of part
	cells	detect stimuli (changes in the environment)
	organs such as brain, spinal cord and pancreas	receive and process information from receptors
	muscles or glands	bring about responses which restore optimum levels

BIOLOGY HIGHER

Main activity: The human nervous system

Time 15 mins

Learning objectives
- To explain how the structure of the nervous system is adapted to its functions
- To describe the structure and function of a reflex arc

Equipment
- mirror
- pencil

1. **Fill in the gaps to describe the nervous system.**

   The nervous system allows humans to react to their surroundings and to c_____ their behaviour. Information about our internal or external environment comes from receptors and passes along cells called _____. The information is in the form of electrical _____. It passes to the central nervous system (CNS). The CNS is made up of the _____ and _____. The CNS coordinates the response of effectors which may be _____ contracting or _____ secreting hormones. Responses known as reflex actions are _____ and rapid. They do not involve the conscious part of the brain. This quick response helps us to avoid _____ to our body.

2. **On the diagram below, draw the structures of a reflex arc that would be involved in moving the hand away from a hot object. Draw in the sensory neurone, relay neurone and motor neurone. Label the stimulus, receptor and effector. Add arrows to show the direction in which the impulse travels.**

   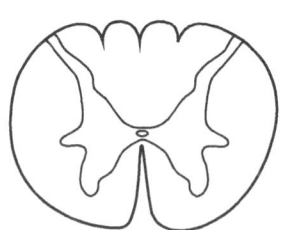

3. **Close and cover your eyes, then open them while looking in a mirror. Explain to your tutor what happens to your iris.**

   This is the reflex response of the eye as it adapts to light. Discuss this with your tutor, then fill in the table below to describe the reflex. Repeat this in the second row for the heat withdrawal reflex considered in question 2.

Stimulus	Receptor	Coordinator	Effector	Response	Purpose of reflex

BIOLOGY HIGHER

Main activity: Thermoregulation   **Time** 15 mins

Learning objectives
- To explain how thermoregulatory mechanisms lower or raise body temperature

Equipment
- pencil

1. Fill in the gaps next to the bullet points and write on the answer lines to complete the diagram which describes the response to changes in core temperature.

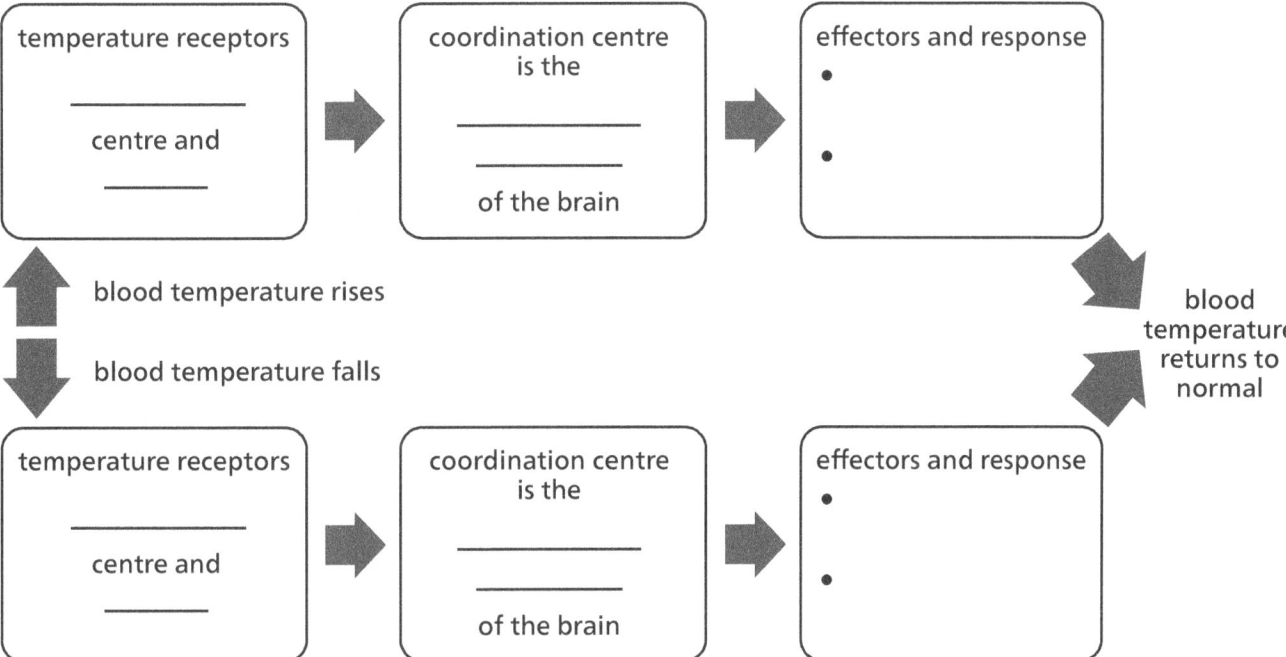

2. Read the following questions and tell your tutor the answers.

   a) What do the arrows between receptor, coordination centre and effectors represent?

   b) By what physical processes can heat be lost from an object?

   c) How do sweating and vasodilation help to cool the body down?

   d) How do shivering and vasoconstriction help body temperature increase to normal levels?

   e) Why would it be misleading to say that vasoconstriction heats the body up?

   f) What is the normal core temperature of the human body?

   g) Thermoregulation is an example of negative feedback. What is negative feedback?

BIOLOGY HIGHER

Homework activity: The reflex arc

Time 15 mins

Learning objectives
- To explain how the structures in a reflex arc are adapted to their function

Equipment
none

1. **The headings in the table below are structures in a reflex arc. Copy the bullet points into the correct column to explain how each structure relates to its function. One column has been completed for you.**

   - connects the sensory and motor neurones
   - chemicals carry the message across the gap allowing information to be passed from neurone to neurone
   - entire cell lies inside the CNS
   - allows coordination of the response
   - forms synapses with motor and sensory neurones
   - carries impulse from CNS to effector
   - forms synapses with effectors
   - gap connecting two neurones
   - elongated cell to carry the impulse through the body

Sensory neurone	Synapse	Relay neurone	Motor neurone
carries impulse from receptors to CNS elongated cell to carry the impulse through the body dendrites form synapses with receptor cells			

2. **The list shows parts of a reflex arc. Write the letters in the correct order in the boxes.**

   A receptor
   B stimulus
   C coordinator
   D response
   E effector

18 Answers

Starter activity: Principles of homeostasis

1. Homeostasis maintains optimal conditions for enzyme action and all cell functions.
2. Nervous: faster response; involves electrical impulses; information sent along cells
   Chemical: longer lasting response; involves hormones released from glands; information sent in the blood plasma
3. Receptors, coordination centres, effectors

Main activity: The human nervous system

1. The nervous system allows humans to react to their surroundings and to *coordinate* their behaviour. Information about our internal or external environment comes from receptors and passes along cells called *neurones*. The information is in the form of electrical *impulses/messages*. It passes to the central nervous system (CNS). The CNS is made up of the *brain* and *spinal cord*. The CNS coordinates the response of effectors which may be *muscles* contracting or *glands* secreting hormones. Responses known as reflex actions are *automatic* and rapid. They do not involve the conscious part of the brain. This quick response helps us to avoid *harm/damage* to our body.
2. Labels should be added as follows: heat stimulus to hand, receptor on hand, sensory neurone leading from hand to CNS, relay neurone within CNS, motor neurone from CNS to muscle in arm, muscle labelled as effector
3.

Stimulus	Receptor	Coordinator	Effector	Response	Purpose of reflex
light	Light receptors on retina	CNS (brain)	Muscles of iris	Pupil becomes smaller/constricts	Protects retina from damage, controls light entering eye
heat	Thermoreceptors/pain receptors on skin	CNS (spinal cord)	Muscles of arm	Muscle contraction moves hand away	Protects skin from damage

Main activity: Thermoregulation

1. Temperature receptors: thermoregulatory centre and skin; coordination centre is the thermoregulatory centre of the brain; body temperature too high: effectors: blood vessels dilate (vasodilation), sweat glands produce sweat; body temperature too low: effectors: blood vessels constrict (vasoconstriction), sweat glands stop sweating, skeletal muscles contract (shiver)
2. a) Nervous impulses
   b) Radiation, convection, conduction, evaporation
   c) They transfer energy from the skin to the environment. When sweating, evaporated water vapour loses energy to the atmosphere. Vasodilation increases the radiation of heat energy away from the skin's surface.
   d) Shivering generates heat by increasing respiration in muscle cells. Vasoconstriction directs blood away from the skin's surface to deeper layers, reducing heat loss by radiation.
   e) It does not generate heat, merely reduces loss of heat from the metabolism.
   f) Approximately 37 °C
   g) Mechanisms which return any changes in the internal environment back to a set point

Homework activity: The reflex arc

Synapse	Relay neurone	Motor neurone
Gap connecting two neurones	Connects the sensory and motor neurones	Carries impulse from CNS to effector
Chemicals carry the message across the gap allowing information to be passed from neurone to neurone	Entire cell lies inside the CNS	Elongated cell to carry the impulse through the body
	Allows coordination of the response	Forms synapses with effectors
	Forms synapses with motor and sensory neurones	

2. B, A, C, E, D

BIOLOGY HIGHER

19 Homeostasis and response: The human brain and eye

Learning objectives

- To identify areas of the brain and describe their functions
- To explain the difficulties of investigating and treating brain damage
- To relate the structures of the eye to their functions
- To describe methods used to treat common eye defects
- To explain how the iris enables adaptation to dim light
- To explain how the eye accommodates to focus on near or distant objects

Specification links

- 4.5.2.2
- 4.5.2.3

Starter activity

- **Brain and eye; 5 minutes; page 126**

  The student should complete the questions on the sheet independently. Go through the answers to check background knowledge on the brain and eye. The third question is developed further in the main activity 'how the eye functions'.

Main activities

- **Studying brain function; 15 minutes; page 127**

  Support the student during the first question. Question 2 a) should be done verbally. For b), draw a line on scrap paper. At one end write: 'beneficial – operate' and at the other end write: 'too risky – don't operate'. Cut out the statements at the bottom of the worksheet in advance and ask the student to position them on the line, explaining why they have placed them there.

- **How the eye functions; 20 minutes; page 128**

  Use the diagram of the eye from the starter activity to aid discussion about question 1. Go through the functions of parts of the eye one by one, adding to the table. For question 2, allow the student to explain the accommodation process initially, then complete the activity one step at a time. Provide support where needed with the diagram.

- **Adapting to dim light; 5 minutes**

  Discuss how the radial and circular muscles work antagonistically to allow the eye to adapt to dim light. Ask the student to sketch a diagram of the iris in bright and dim light showing these muscles either contracted or relaxed. Remind the student that this is a simple reflex action.

Plenary activity

- **Traffic lights; 5 minutes**

  Provide the student with red/orange/green coloured pencils or highlighters. Look back over each part of the lesson and ask the student to mark each part with a circle. Red means 'don't understand', amber means 'mostly understand' and green means 'fully understand'.

Homework activity

- **The eye – exam-style questions; 15 minutes; page 129**

  Ask the student to complete the exam-style questions about the adaptations of the eye in dim light.

Support ideas

- **Studying brain function** Show photos from textbooks or the internet of brain images acquired by MRI and fMRI.
- **How the eye functions** Use a partly inflated balloon or small flexible ball to show how pressure applied in a ring around the outside distorts the shape.

Extension ideas

- **Studying brain function** Discuss some famous cases of brain damage that provided early understanding of brain function, such as Phineas Gage and Louis Leborgne.
- **How the eye functions** Ask the student to draw light ray diagrams to show how an image is inverted on the retina.

Progress and observations

BIOLOGY HIGHER

Starter activity: Brain and eye

Time 5 mins

Learning objectives
- To know the basic structure of the brain
- To identify the structures of the eye

Equipment
none

1. Write down three words that best describe what the brain does.

_____

2. What is the brain made of?

_____

3. In the table, write the names of structures of the eye to match the descriptions. Then find the structure on the diagram and write the letter label in the first column. One has been done for you.

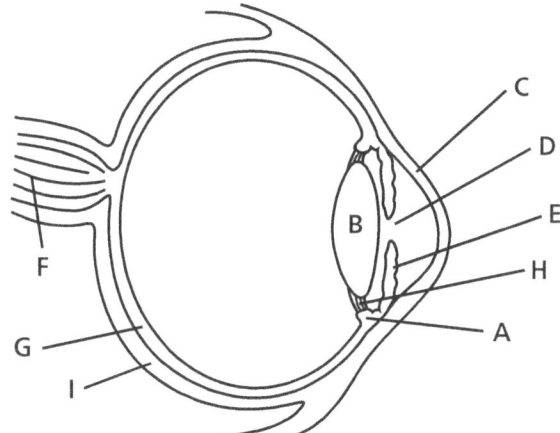

Letter	Structure	Description and function
C	cornea	The transparent, curved front of the eye helps to bend light rays.
		They contract and relax, causing the lens to change shape when focusing on an object.
		It holds the lens in place.
		It consists of an inner ring of circular muscle and an outer layer of radial muscle. Its function is to control the amount of light entering the eye.
		It contains sensory neurones. Carries the impulses from the receptors to the brain.
		A hole in the middle of the iris where light passes through. In bright light it is constricted and in dim light it is dilated.
		A transparent, flexible, curved structure. Focuses incoming light rays onto the retina.
		A layer of photoreceptors (rod and cone cells), which respond to light intensity and colour.
		A tough protective layer on the outside of the eyeball, continuous with the cornea.

 **BIOLOGY HIGHER**

Main activity: Studying brain function

Time 15 mins

Learning objectives

- To identify areas of the brain and describe their functions
- To explain some of the difficulties of investigating brain function and treating brain damage and disease

Equipment

- cut out statements from the table
- benefits/risk scale

1. **The brain is a complex organ. It has different regions that carry out different functions. Label the brain diagram with the names of regions listed in the table. Then write down the function of each region.**

Region of brain	Function
cerebral cortex	
cerebellum	
medulla	

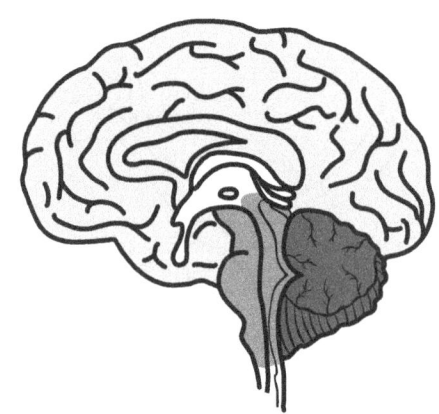

2. **A patient has been diagnosed with a tumour in their brain.**

   a) Explain to your tutor how neuroscientists have mapped the regions of the brain to particular functions.

   b) Your tutor will provide you with cut-outs of the statements below and a scale of benefits/risk. Place the statements on the line to evaluate whether the doctor should decide to operate and remove the tumour or not.

The brain is a delicate organ and is very easily damaged.	The tumour is in an area associated with hearing.	The patient has lost feeling and movement in one side of the face.
The tumour is quite deep in the brain.	The tumour is growing very slowly.	Brain tissue does not fully repair itself after damage.
The tumour is malignant.	An MRI scan has shown that the tumour is small.	The patient has other health issues and difficulties with breathing.

BIOLOGY HIGHER

Main activity: How the eye functions

Time 20 mins

Learning objectives
- To relate the structures of the eye to their functions
- To explain how the eye accommodates to focus on near or distant objects
- To describe methods used to treat common eye defects

Equipment
- pencil
- spare paper

1. The ciliary muscles contract or relax to change the shape of the lens and focus the light from near or distant objects onto the retina. This process of focusing is called accommodation. Complete the second diagram and cross out words to finish the sentences and explain how the eye accommodates to focus on near and distant objects.

Near object
The lens is thick/thin.
The ciliary muscles are contracted/relaxed.
The sensory ligaments are pulled tight/loose.
The light is refracted a lot/a little.

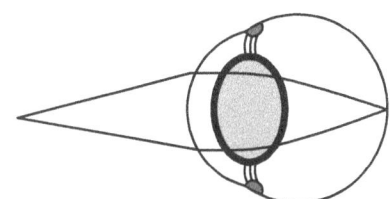

Far object
The lens is thick/thin.
The ciliary muscles are contracted/relaxed.
The sensory ligaments are pulled tight/loose.
The light is refracted a lot/a little.

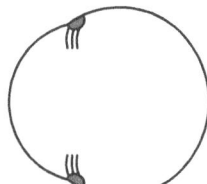

2. Two common defects of the eyes are myopia (short sightedness) and hyperopia (long sightedness); decide which is represented by each of the diagrams below and label it accordingly.

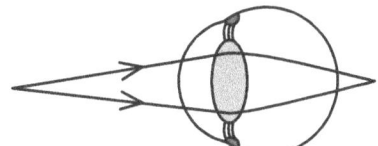   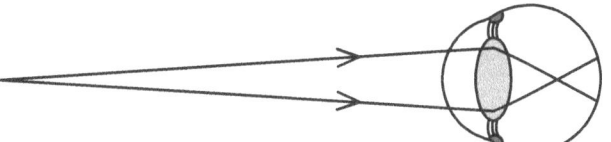

3. These defects are usually treated with spectacle lenses which refract the light rays so that they do focus on the retina. On a separate piece of paper, redraw these ray diagrams to show how a spectacle lens can be used to correct each condition.

4. Discuss with your tutor what new technologies are now available to treat myopia and hyperopia.

BIOLOGY HIGHER

Homework activity: The eye – exam-style questions

Time 15 mins

Learning objectives
- To explain how the iris enables adaptation to dim light

Equipment
none

1. The diagram shows the inside of an eye.

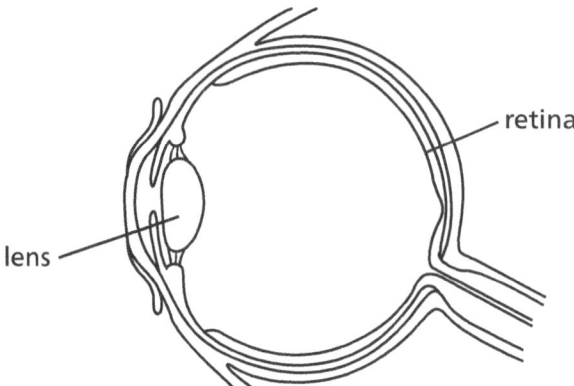

a) Describe the job of the lens.

_____
_____

b) The optic nerve carries impulses to the brain. Label the optic nerve on the diagram above.

c) The impulses travel along neurones in the optic nerve. What type of neurone carries impulses to the brain?

_____

2. A boy walks from a brightly lit room into a darker room. His eyes adjust so that he can still see clearly.

a) This is caused by changes in the iris muscles. Describe the changes in these muscles.

_____

b) Explain how the changes help the boy to see in the darker room.

_____

3. The eye is a sense organ. It has receptors in the retina. Name two things that these receptors are sensitive to.

_____ and _____

BIOLOGY HIGHER

19 Answers

Starter activity: Brain and eye

1. Student's own answers. Examples include control, complex behaviour, coordination.
2. Large numbers/billions of interconnected neurones
3. Order going down the table: C: cornea, A: ciliary muscle, H: suspensory ligaments, E: iris, F: optic nerve, D: pupil, B: lens, G: retina, I: sclera

Main activity: Studying brain function

1. Cerebral cortex controls complex behaviour; thought, memory, reasoning, interpretation of sensory information Cerebellum: coordination of balance and complex movements
Medulla: controls automatic actions like changes in heartbeat, blood pressure and breathing
2. a) By studying patients with brain damage and noting the effects; by electrically stimulating different parts of the brain and getting the patient to describe what they feel or observing the effects; by using MRI scanning techniques in real time, where areas of the brain that are active and using more oxygen 'light up' on the scans while doing certain tasks (fMRI)
   b) There are no right answers here. Arguments for treatment might be that the tumour appears to be affecting other parts of the brain and that it might spread. Arguments against treatment might be those statements that indicate a risk of brain damage and that effects on hearing might be less severe than tumours in areas controlling key bodily functions.

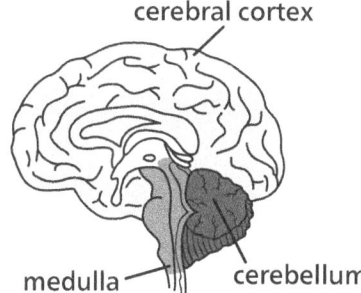

Main activity: How the eye functions

1. Near object: the lens is thick; the ciliary muscles are contracted; the sensory ligaments are loose; the light is refracted a lot; far object: the lens is thin; the ciliary muscles are relaxed; the sensory ligaments are pulled tight; the light is refracted a little. Diagram of far object should show a narrower lens stretched from top to bottom with light rays coming from a more distant point, or arriving as parallel rays, being bent slightly by the cornea but mainly by the lens to focus on the retina
2. Left diagram is hyperopia; right diagram is myopia
3. Redrawn diagram for hyperopia should show a converging lens which refracts the light inwards before it reaches the eye; redrawn myopia should show a diverging (concave) lens which bends the light rays outwards before they reach the eye
4. Hard and soft contact lenses; laser surgery to change the shape of the cornea; replacement lens in the eye

Homework activity: The eye – exam-style questions

1. a) To focus light/form a clear image onto the retina
   b) Ensure the optic nerve is clearly labelled
   c) sensory
2. a) Circular relax; radial contract
   b) Pupil is enlarged so more light enters the eye
3. Light intensity; colour of light/wavelength

20 Homeostasis and response: Hormonal coordination in humans

Learning objectives

- To describe the role of hormones and the endocrine system
- To locate and explain the function of endocrine glands
- To explain how blood glucose levels are controlled in the body
- To compare Type 1 and Type 2 diabetes and their treatment

Specification links

- 4.5.3.1
- 4.5.3.2
- MS 1c

Starter activity

- **Human endocrine system; 10 minutes; page 132**

  This activity covers basic background knowledge about the human endocrine system. Ask the student to complete the activity independently. Check their answers and review any areas of uncertainty.

Main activities

- **Control of blood glucose concentration; 15 minutes; page 133**

  Start by reviewing why blood glucose levels must be maintained: some blood glucose is required at all times to supply to body cells for respiration, but too much can have osmotic effects, causing cell damage. Mention the serious effects of diabetes. Ask the student to recall the two hormones involved, then ask them to complete the flow diagram in question 1. Question 2 considers the process of glucose uptake and release at the cellular level in more detail. Allow the student time to come up with their own ideas here.

- **Type 1 and Type 2 diabetes; 15 minutes; page 134**

  Allow the student to complete question one independently. In question 2, use the graph as a focus for verbal discussion of questions a) to d) with annotation of the graph where appropriate.

- **Hormones versus nerve impulses; 5 minutes**

  Ask the student to list the similarities and differences between the two methods of communication and control. Emphasise that compared with the nervous system hormone effects are slower but act for longer.

Plenary activity

- **Two stars and a wish; 5 minutes**

  Ask the student to tell you two things that they have enjoyed, done well or learned in the lesson and one question or target for further learning.

Homework activity

- **Obesity and diabetes; 20 minutes; page 135**

  This activity is an extended exam-style question to evaluate the link between obesity and Type 2 diabetes.

Support ideas

- **Control of blood glucose concentration** Show a short video animation to show the relationship between insulin and glucose. An internet search for 'glucose cell uptake video' will give several useful results.
- **Type 1 and Type 2 diabetes** Discuss in more detail the more serious long-term effects of diabetes. Ask the student to explain why raised blood glucose might result in problems such as blindness and nerve damage.

Extension ideas

- **Human endocrine system** Consider the pituitary gland as a master gland in more detail. Ask the student to name examples of pituitary hormones which act on other glands to stimulate other hormones to bring about effects.
- **Type 1 and Type 2 diabetes** Sketch or show a drawing to explain how insulin binds to receptors on the cell surface and how this results in increased uptake of glucose.

Progress and observations

BIOLOGY HIGHER

Starter activity: Human endocrine system   Time 10 mins

Learning objectives
- To describe the endocrine system
- To identify the position of endocrine glands on a diagram of the human body

Equipment
none

1. **Delete the incorrect words to complete the sentences about the endocrine system.**

    The endocrine system is composed of muscles/glands/organs which secrete chemicals called hormones/enzymes/antibodies directly into the bloodstream. The blood carries the hormone to a target/nearby/sense organ where it produces an effect.

2. **The diagram shows the position of glands in the human endocrine system (both male and female). Write the letter that shows the position of each gland next to its name on the table. Then draw a line to match this gland to the hormones it produces.**

Letter	Gland	Hormones produced
	pituitary gland	thyroid hormones: control chemical reactions in the body
	pancreas	adrenaline: increases heart and breathing rate for 'fight or flight'
	thyroid	a 'master gland': produces hormones that control other glands
	adrenal glands	testosterone: male body changes at puberty
	ovaries	insulin and glucagon: control blood sugar
	testes	oestrogen and progesterone: female body changes at puberty and control of the menstrual cycle

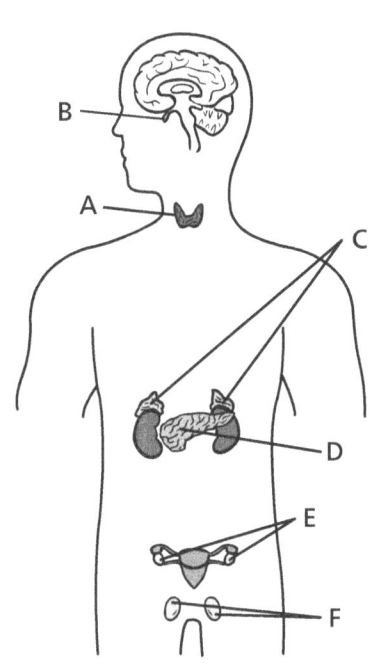

132

BIOLOGY HIGHER

Main activity: Control of blood glucose concentration

Time 15 mins

Learning objectives
- To explain how glucagon interacts with insulin in a negative feedback cycle to control blood glucose levels in the body

Equipment
- pencil

1. In heathy people blood glucose concentrations are kept within a narrow range by the interaction between two hormones. Complete the missing words in the flow diagram below to summarise this process.

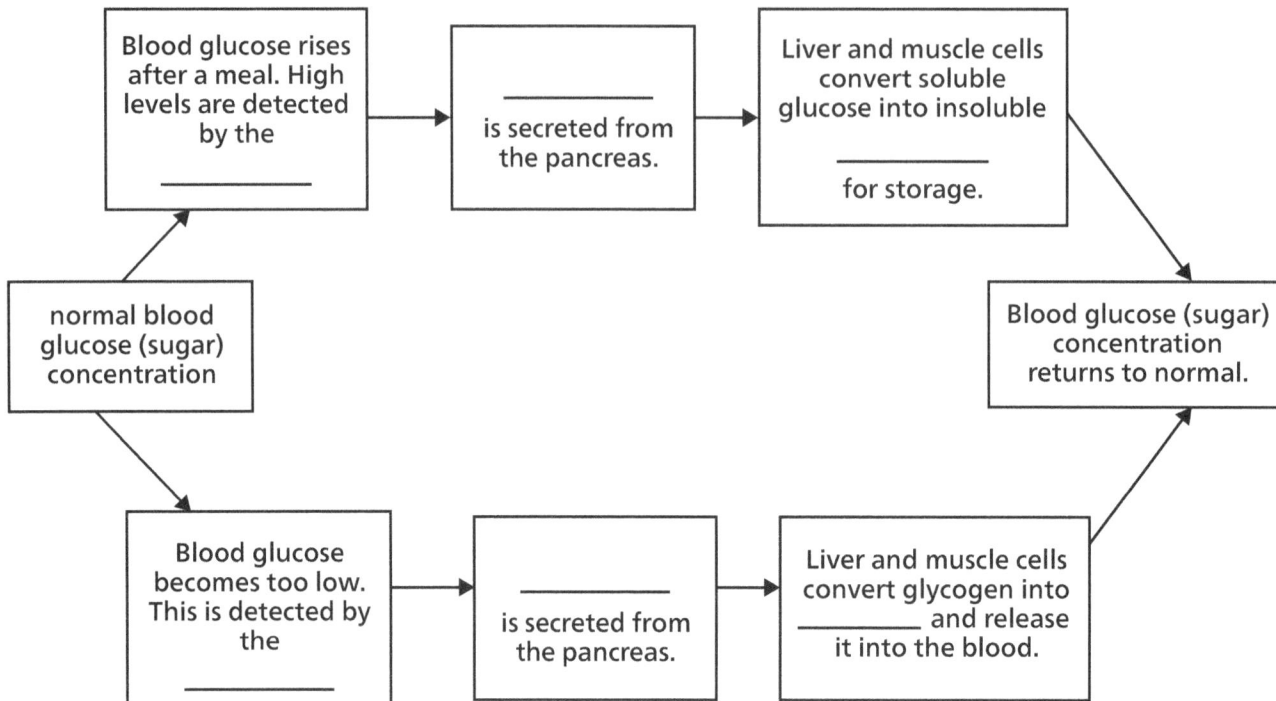

2. In the process shown above, if glucose concentrations rise or fall, they are brought back to a certain set-point. What is the name of this type of process?

_____

3. Add labels and arrows to the boxes below to show in detail what happens in liver and muscle cells when the blood sugar concentration is too high or too low.

Blood glucose too high	Blood glucose too low
pancreas	pancreas
blood	blood
liver and muscle cells	liver and muscle cells

BIOLOGY HIGHER

Main activity: Type 1 and Type 2 diabetes Time 15 mins

Learning objectives
- To compare Type 1 and Type 2 diabetes and explain how they can be treated

Equipment
- pencil

1. Complete the table to compare Type 1 and Type 2 diabetes.

Type	Problem	Treatment
	The pancreas does not produce enough insulin. The person has uncontrolled high blood glucose levels.	
	Insulin is produced by the pancreas but body cells no longer respond to it. Blood glucose levels remain high after a meal.	

2. The graph shows the blood glucose levels of one person with diabetes and one person who does not suffer from the condition. Their blood glucose levels were monitored for eight hours after they had eaten the same meal containing carbohydrates.

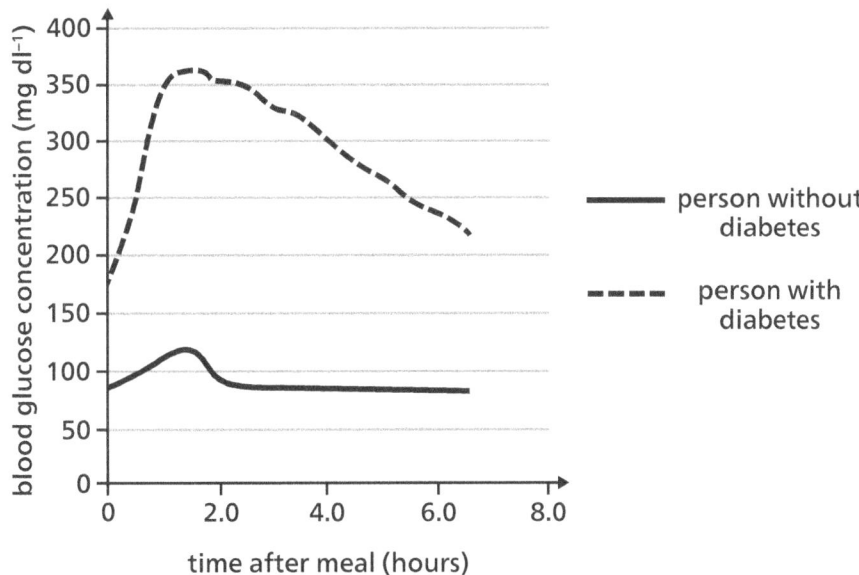

Discuss the following points with your tutor:

a) Why do blood glucose levels take more than an hour to reach their maximum in both people?

b) Why do the non-diabetic person's blood sugar levels not rise very much?

c) The person with diabetes suffers from Type 1 diabetes where very little, if any, insulin is produced. Why does the blood glucose concentration still decline?

d) Sketch a line on the graph to show what blood glucose levels would look like in the person with diabetes if they had injected insulin just before their meal.

BIOLOGY HIGHER

Homework activity: Obesity and diabetes   Time 20 mins

Learning objectives
- To evaluate information on the relationship between obesity and diabetes

Equipment
none

A report called 'Adult obesity and Type 2 diabetes' published by Public Health England gave the following evidence about the link between Type 2 diabetes and obesity.

- The number of people with obesity is rising in England. The number of people with diabetes is also rising.
- 62% of adults were overweight or obese in England in 2012.
- 6% of people aged 17 years or older had been diagnosed with diabetes in England in 2013.
- 90% of adults with Type 2 diabetes aged 16–54 years are overweight or obese in England.
- 12.4% of people aged 18 years and over with obesity have diabetes, five times that of people with a healthy weight.

Source: Gatineau M, Hancock C, Holman N, Outhwaite H, Oldridge L, Christie A, Ells L. Adult obesity and type 2 diabetes. Oxford: Public Health England, 2014.

---

1. **Evaluate the evidence that obesity may be linked to Type 2 diabetes. Use the information provided to support your answer.**

[6 marks]

BIOLOGY HIGHER

20 Answers

Starter activity: Human endocrine system

1. The endocrine system is composed of *glands* which secrete chemicals called *hormones* directly into the bloodstream. The blood carries the hormone to a *target* organ where it produces an effect.
2. B: Pituitary gland, a 'master gland': produces hormones that control other glands
   D: Pancreas; insulin and glucagon: control blood sugar
   A: Thyroid; thyroid hormones: control chemical reactions in the body
   C: Adrenal glands; adrenaline: increases heart and breathing rate for 'fight or flight'
   E: Ovaries; oestrogen and progesterone: female body changes at puberty and control of the menstrual cycle
   F: Testes; testosterone: male body changes at puberty

Main activity: Control of blood glucose concentration

1. Top row: pancreas, insulin, glycogen
   Bottom row: pancreas, glucagon, glucose
2. Negative feedback
3. Arrows and labels should show that where blood glucose concentration is too high, the pancreas releases the hormone insulin into the blood and that this causes glucose to move from the blood into the cells, where excess glucose is converted to glycogen.
   Where blood glucose concentration is too low, the diagram should show that the pancreas releases the hormone glucagon into the blood and that this that causes glycogen to be converted into glucose and released into the blood.

Main activity: Type 1 and Type 2 diabetes

1. Type 1: insulin injections; Type 2: carbohydrate-controlled diet and an exercise regime
2. a) Food must be digested before it enters the bloodstream.
   b) Insulin from the pancreas causes increased uptake of glucose into body cells.
   c) Blood glucose is lost via the urine in people with diabetes; a small amount will be used in respiration.
   d) Glucose peak and final glucose concentration will be lower.

Homework activity: Obesity and diabetes

1. The following table provides guidance on what a Level 3, 2 or 1 answer to this question would look like and the number of marks each would attract.

L3	A detailed, balanced consideration of the evidence, with relevant use made of statements. Summarises the strength of the relationship effectively, but includes clear arguments why this alone does not indicate that obesity causes type 2 diabetes.	5–6 marks
L2	An attempt to evaluate evidence is made. Relevant use is made of statements but the argument may lack balance. It is recognised that a relationship does not prove cause but lacks detailed explanation.	3–4 marks
L1	Discrete relevant points made. The strength of the relationship is summarised using some evidence but there is no balance to the argument and causation may be implied.	1–2 marks
	Indicative content • The correlation may be just down to chance. • A third factor may link both obesity and diabetes. • A causal mechanism must be found to prove cause and effect. • The proportion of people with diabetes that are overweight or obese is much higher (90%) than that of the general population (62%). • A person with Type 2 diabetes is 28% more likely to be overweight or obese. • Proportion of population with diabetes is quite low (6%), but proportion that are overweight or obese is high (62%). • So most overweight or obese people do not suffer from Type 2 diabetes. • But obesity is a risk factor for Type 2 diabetes because obese adults are five times more likely to have the condition.	

BIOLOGY HIGHER

21 Homeostasis and response: Water and nitrogen balance

Learning objectives

- To explain the effect on cells of osmotic changes in body fluids
- To describe the function of kidneys and ADH in maintaining water balance
- To describe how kidney dialysis works
- To understand how excess amino acids are removed from the body

Specification links

- 4.5.3.3

Starter activity

- **Water losses and gains; 5 minutes; page 138**

    For question 1, osmosis should be mentioned. For question 2, discuss the importance of each route, and emphasise that sweating, exhalation and urine are the most important routes for water loss. Emphasise that the body has no control over the loss of water, ion or urea through the skin.

Main activities

- **Maintaining water balance; 15 minutes; page 139**

    First, use the diagram to describe how the blood is filtered to produce urine. Ask the student to complete question 1. Question 2 covers the main specification knowledge, but also ask about large proteins and blood cells as examples of items too large to pass through. Ask the student to explain how ADH controls water balance, then complete the gaps in the flow diagram. Remind the student that this is negative feedback.

- **Nitrogen balance and dialysis; 15 minutes; page 140**

    The student should construct the flow diagram on a separate piece of paper. Question 2 provides the basis for a verbal discussion about dialysis treatment. The diagram could be annotated. This is developed further in the next activity.

- **Treating organ failure; 10 minutes**

    With the focus on dialysis versus kidney transplant, discuss the advantages and disadvantages of treating organ failure by mechanical device or transplant. Ask the student to record their ideas on a spider diagram. This is also covered by the homework question.

Plenary activity

- **Homeostasis awards; 5 minutes**

    Pancreas, pituitary, liver or kidney – which organ should get the award for 'services to homeostasis'? Ask the student to verbally put a brief case for each. Look for misconceptions about the role of the liver and kidney in nitrogen balance.

Homework activity

- **Dialysis and kidney transplants; 20 minutes; page 141**

    This activity has exam-style questions to compare dialysis and kidney transplants.

Support ideas

- **Maintaining water balance** Model the idea of a partially permeable membrane using a sieve and a mixture of gravel and flour in water. Emphasise that filtration in the kidney is also under pressure.
- **Nitrogen balance and dialysis** If the student struggles to produce the flow diagram in question 1, provide them with a highlighter to highlight keywords that should be included as a first step.

Extension ideas

- **Maintaining water balance** Discuss the effects of other intakes into the body on ADH release and water balance, for example ecstasy (ADH release increased) and alcohol (ADH release blocked); link this to harmful effects.
- **Nitrogen balance and dialysis** Draw a molecular diagram of an amino acid and ask the student to identify which part might be removed in deamination.

Progress and observations

Starter activity: Water losses and gains

Time 5 mins

Learning objectives
- To explain the effect on cells of osmotic changes in body fluids
- To describe losses of substances through the skin or lungs

Equipment
none

1. Control of water balance in the body keeps the concentration of the blood constant. Give two reasons why this is important.

_____

_____

2. Place ticks in the table to show how water in the body can be lost or gained. One has been done for you.

	Water lost	Water gained
drinks		
exhalation		
faeces		
food		
respiration		
sweating	✓	
urine		

3. Name two other substances that can be lost through the skin in sweat.

_____

_____

138

BIOLOGY HIGHER

Main activity: Maintaining water balance   **Time** 15 mins

Learning objectives

- To describe the function of the kidneys in maintaining the water balance of the body

Equipment

none

1. The kidneys are important organs. They control the water balance of the body. They also allow excess ions (salt) and the waste product urea to be removed from the body in urine. The kidneys produce urine by filtration of the blood in the kidney tubules. The diagram shows a kidney tubule. Match the letters on the diagram to the step in urine production in the table.

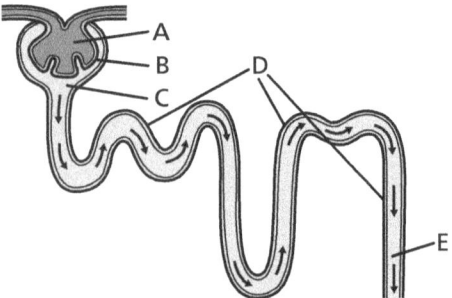

Letter	Step in urine production
	filtration of blood through a partially permeable membrane
	urine passes out of the kidney tubule
	blood under high pressure
	selective reabsorption of ions, glucose and water back into the blood
	only small molecules pass through into filtrate

2. Only small molecules like water, ions, glucose and urea can pass from the blood into the kidney tubule. Decide what happens to each of these types of molecules.

   These molecules are all reabsorbed back into the blood. _____

   Excess molecules pass into the urine, the rest are reabsorbed. _____ and _____

   There is no reabsorption. _____

3. The water level in the body is controlled by the hormone ADH which acts on the kidney tubules. Complete the flow diagram to show how the release of more or less ADH controls water balance.

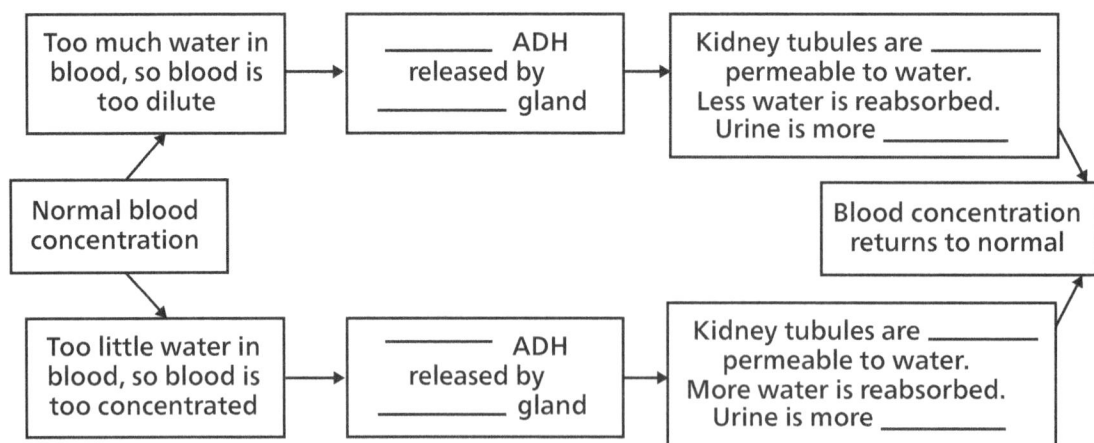

4. What behavioural response might be prompted by having too little water in the blood?

   _____

139

BIOLOGY HIGHER

Main activity: Nitrogen balance and dialysis

Time 15 mins

Learning objectives
- To describe how excess amino acids are converted into ammonia
- To describe how kidney dialysis works

Equipment
- pencil
- spare paper

1. Read the paragraph about how nitrogen balance is controlled by the process of deamination. Use information in the paragraph to make a flow diagram on a separate piece of paper, explaining how excess amino acids are dealt with by the body.

> **Deamination**
> When we eat and digest proteins in our food we may often take in more amino acids than we need. We cannot store them but it would be wasteful to excrete them without getting some useful energy from them. The excess amino acids are first deaminated in the liver. The part of the amino acid containing nitrogen is removed and converted into ammonia. The energy-rich part of the molecule that remains is modified and stored as carbohydrate or fat. Ammonia is toxic and so it is immediately converted to urea. Urea is a much less toxic molecule and can be carried in the blood plasma to the kidneys, where it is filtered out and safely excreted.

2. If a person's kidneys do not work properly, they may need dialysis to remove urea and other waste products from the blood. The diagram shows how dialysis works.

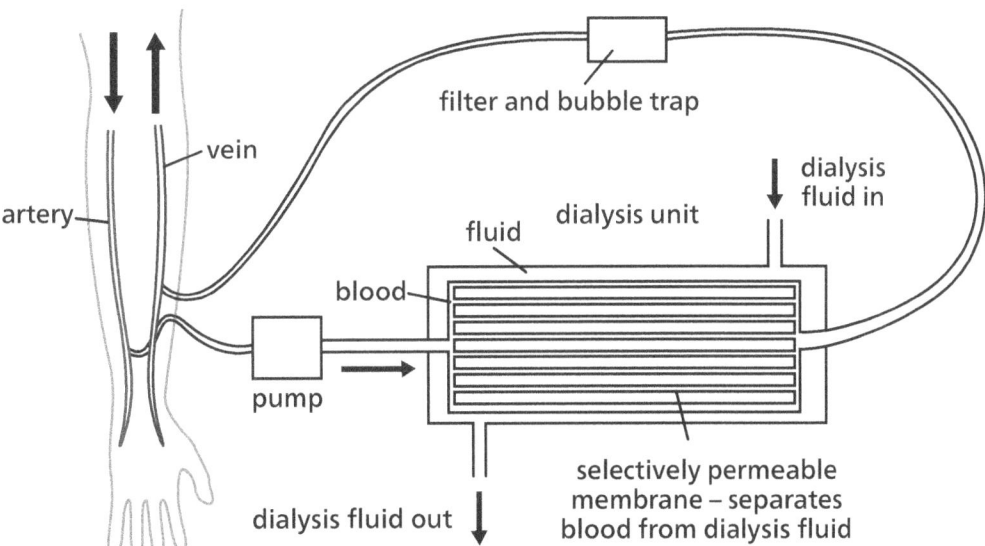

Explain the following to your tutor:

a) What does the pump do?

b) What happens at the partially permeable membrane?

c) How is the waste from the blood removed?

d) How is the loss of useful substances from the blood, such as glucose, prevented?

140

BIOLOGY HIGHER

Homework activity: Dialysis and kidney transplants

Time 20 mins

Learning objectives
- To compare treatment of organ failure with dialysis and kidney transplant

Equipment
none

1. **During dialysis, urea passes out of the blood into the dialysis fluid by diffusion.**

   a) Explain why urea diffuses out of the blood into the dialysis fluid.

   b) Why is it important to maintain balanced water levels in cells in the human body?

   c) Answer this exam-style question.

   > People who suffer from kidney failure may be treated by organ transplant or by using kidney dialysis. Compare the advantages and disadvantages of these two methods of treatment.
   >
   > [6 marks]

21 Answers

Starter activity: Water losses and gains

1. Affects osmosis in cells; changes concentration in the cell; this affects cell reactions or enzymes; cells do not function efficiently; cells shrink, expand or burst
2. Water gained: drinks, food, respiration; water lost: exhalation, sweating, urine, faeces
3. Ions and urea

Main activity: Maintaining water balance

1. B, E, A, D, C
2. Glucose; ions and water; urea
3. Too much water (top row); less (ADH), pituitary, less (permeable), (more) dilute; too little water (bottom row); more (ADH), pituitary, more (permeable), (more) concentrated
4. Drinking fluid in response to feeling thirsty

Main activity: Nitrogen balance and dialysis

1. The flow diagram should include clear detail of deamination to ammonia and then conversion to less toxic urea. It should be clear that this takes place in the liver.
2. a) Moves the blood between the membranes
   b) Small ions can pass through the membrane by diffusion. The net movement is down the concentration gradient.
   c) Waste such as urea and excess ions diffuse down the concentration gradient into the dialysis fluid, which flows through the machine, taking it away. It is replaced by clean fluid.
   d) Useful substances must be at the same concentration in the dialysis fluid as in the blood, so that there will be no concentration gradient and no net movement.

Homework activity: Dialysis and kidney transplants

1. a) Particles or molecules move; from high to low concentration; (urea) concentration high in blood/(urea) concentration low in dialysis fluid
   b) Any two from: water balance affects osmosis; affects movement of substances in and out of cells; affects cell reactions/efficiency of cell function/enzymes; cell contents become more dilute or more concentrated; cells may shrink/expand/burst
   c) The following table provides guidance on what a Level 3, 2 or 1 answer to this question would look like and the number of marks each would attract.

L3	A detailed and coherent evaluation is provided that considers a number of advantages and disadvantages for each treatment and comes to an overall conclusion consistent with the reasoning.	5–6 marks
L2	An attempt to describe both advantages and disadvantages for both treatments. Arguments may be less coherent for one treatment.	3–4 marks
L1	Some relevant points made that are clearly advantages or disadvantages but these may be limited to one treatment or lack balance. The logic may be unclear and the conclusion, if present, may not be consistent with the reasoning.	1–2 marks
	Indicative content Disadvantages • Dialysis needs a lot of time attached to machine • Dialysis needs long-term regular treatment • Transplant carries risk of rejection • Delay/may be difficult to find suitable donor • Regular dialysis carries long-term risk of infection • Transplant has risks linked to major surgery • Transplant involves need to take immune-suppressant drugs for life • Transplant carries increased risk of infections as a result of immune-suppressant	Advantages • Transplant allows normal lifestyle, no time commitment • Transplant may provide permanent cure • Transplant may be cheaper in the long term • Dialysis is readily available	

22 Homeostasis and response: Human reproduction and contraception

Learning objectives

- To describe the roles of hormones in human reproduction
- To explain how hormones interact in the control of the menstrual cycle
- To evaluate hormonal and non-hormonal methods of contraception

Specification links

- 4.5.3.4
- 4.5.3.5

Starter activity

- **Male and female reproductive hormones; 5 minutes; page 144**

  Ask the student to record their current knowledge about human reproductive hormones as a spider diagram using the starting point provided. This is best done as a timed activity. The student's work should not be corrected or added to because this activity will be revisited and added to in the plenary.

Main activities

- **Hormones and the menstrual cycle; 15 minutes; page 145**

  Ask the student to complete the first two columns of the table independently, then discuss the interactions between hormones and add arrows to the final column. Work through question 2, one section at a time, helping the student to annotate the diagram as indicated, using pencil so that any errors can be corrected.

- **Contraception; 20 minutes; page 146**

  Do not provide the student with the activity sheet initially. Provide the student with a cut out version of the cards with hormone names and descriptions separated. Read the instructions for the various sorting activities and support them as required. For parts c) and d) give the student the activity sheet so that they can annotate it if they want to.

- **Secondary sexual characteristics; 5 minutes**

  Ask the student to recall and verbally list the male and female secondary sexual characteristics that are influenced by testosterone and oestrogen.

Plenary activity

- **Male and female hormones revisited; 5 minutes**

  Give the student a short, timed period to add what they can to the spider diagram on the starter worksheet. Discuss any remaining areas of uncertainty. Make sure that the action of reproductive hormones during puberty causing secondary sexual characteristics to develop is included. The action of testosterone should also be included as the main male reproductive hormone, produced by the testes and stimulating sperm production.

Homework activity

- **Reproduction and hormones quick quiz; 20 minutes; page 147**

  This activity has a series of short answer questions about reproductive hormones and contraception.

Support ideas

- **Hormones and the menstrual cycle** The complex interactions between hormones can make it more difficult to understand their main effects. Get the student to highlight just the main effects (on egg development and uterus lining) and focus on this initially.
- **Contraception** The student may be unfamiliar with some methods. It may be necessary to show diagrams or pictures to aid understanding.

Extension ideas

- **Hormones and the menstrual cycle** Discuss the involvement of the developing follicle and its decline on progesterone release. Show pictures of micrographs of follicles. A web search for 'ovarian follicle photos' will give suitable results.
- **Contraception** Ask the student to suggest the types of people who would be advised against using a) sterilisation (young people who haven't had children) and b) oral contraceptives (anyone at risk of DVT, smokers, etc.)

Progress and observations

BIOLOGY HIGHER

Starter activity: Male and female reproductive hormones Time 5 mins

Learning objectives
- To describe the roles of human reproductive hormones

Equipment
- pencil

1. Fill in and add to this spider diagram about human reproductive hormones and their role in the body. You should be able to list the names of one male and four female reproductive hormones.

   Then add to the diagram, for example, giving information on where they are produced and the effects of these hormones.

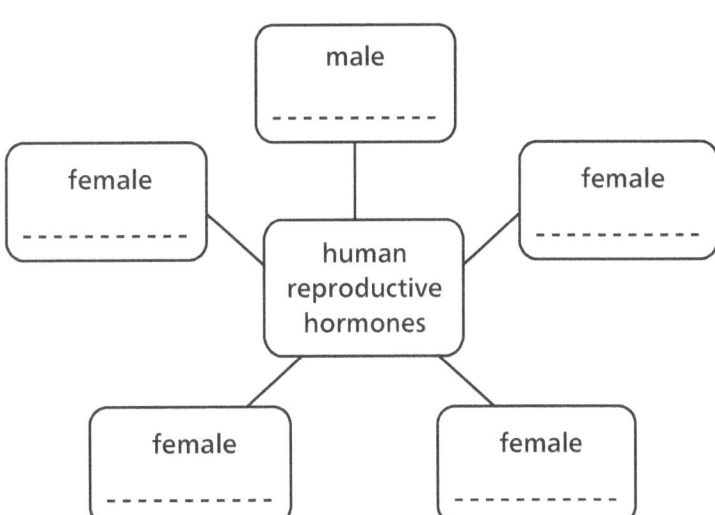

BIOLOGY HIGHER

Main activity: Hormones and the menstrual cycle Time 15 mins

Learning objectives
- To explain the interactions of FSH, oestrogen, LH and progesterone in the control of the menstrual cycle

Equipment
- coloured pencils

1. Fill in the gaps in the table to name the female hormone that matches each function, and where it is produced. Add arrows in the 'interactions' column to show where one hormone acts on another. Add a positive sign if the hormone stimulates the release of another hormone. Add a negative sign if it inhibits the release of the second hormone.

Hormone	Produced	Function	Interactions
		• causes an egg to mature in an ovary • stimulates the ovaries to release oestrogen	
		• stops FSH being produced so only one egg matures per cycle • stimulates the pituitary gland to release LH • causes the lining of the womb to thicken and be maintained	
		• stimulates the release of the mature egg from the ovary	
		• maintains the thickened lining of the uterus (stays high during pregnancy) • inhibits the release of both FSH and LH	

2. The diagram shows changes in two hormones and in the thickness of the lining of the uterus during the human menstrual cycle.

   a) Add another line to show how the levels of LH change during the cycle. Explain to your tutor how you know where the peak in LH will be on the graph.

   b) Add these labels to the diagram: FSH is high – egg is maturing; menstruation; uterus lining builds up; uterus lining maintained

   c) Add labelled lines to the diagram to show what would happen to the progesterone levels and uterus lining if the woman became pregnant.

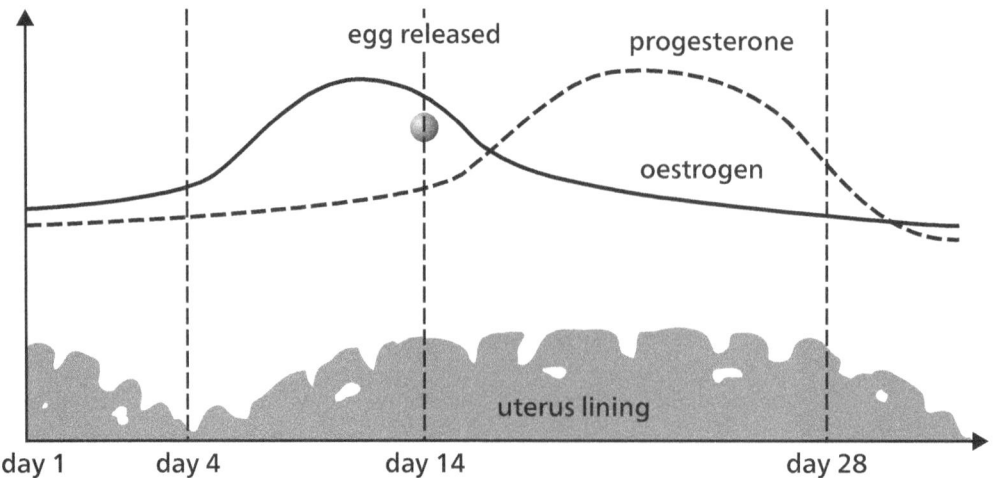

145

BIOLOGY HIGHER

Main activity: Contraception

Time 20 mins

Learning objectives

- To evaluate the different hormonal and non-hormonal methods of contraception

Equipment

- cut out set of the boxes below to make sort cards

1. Fertility can be controlled by a variety of hormonal and non-hormonal methods of contraception. The shaded boxes below contain the names of the methods. The other boxes contain a description including how the method works. Your tutor will provide you with a cut out version of the boxes.

   a) First, sort the method names (grey boxes) into hormonal and non-hormonal methods.

   b) Then match the description boxes to the method names.

   c) Then order the methods from most reliable to least reliable in terms of preventing pregnancy.

   d) Finally, for each method tell your tutor one advantage and one disadvantage of that form of contraception.

oral contraceptives	contain progesterone and oestrogen to inhibit FSH production so that no eggs mature; must be taken every day, may cause some side effects	**spermicidal agents**	kill or disable sperm, for example spermicidal gel placed in the vagina
injection, implant or skin patch of slow release progesterone	inhibits the maturation and release of eggs for a number of months or years, may cause some side effects	**abstaining from intercourse when an egg may be in the oviduct**	sperm and egg are less likely to come into contact during the critical time for fertilisation
barrier methods such as condoms and diaphragms	prevent the sperm reaching an egg; condoms also prevent STIs	**surgical methods of male and female sterilisation**	cause permanent infertility and are very difficult to reverse; small health risk from medical procedure
intrauterine devices	prevent the implantation of an embryo or release a hormone that prevents fertilisation; must be inserted by a doctor, slight risk of infection		

BIOLOGY HIGHER

Homework activity: Reproduction and hormones quick quiz

Time 20 mins

Learning objectives
- To describe the role of hormones in human reproduction and contraception

Equipment
none

Answer the following quick-quiz questions.

1. What is the purpose of the thickened womb lining?

2. At what time in a woman's life do eggs begin to mature?

3. What is the name of the release of an egg from the ovary?

4. How often does this egg release happen in most women?

5. Where are the target cells for FSH and LH?

6. How do these hormones reach their target cells?

7. Give two roles of testosterone.

8. Name two male secondary sexual characteristics.

9. What contraceptive method involves closing or blocking the fallopian tubes?

10. What hormone do contraceptive implants contain?

22 Answers

Starter activity: Male and female reproductive hormones

1. Basic detail: female hormones: FSH, oestrogen, LH and progesterone; male hormone: testosterone, produced by the testes, stimulates sperm production; oestrogen and testosterone noted as the main female and male hormones, respectively, that are associated with development of secondary sexual characteristics
Further details should include where the hormone is produced and basic ideas about function.

Main activity: Hormones and the menstrual cycle

1. In order going down the table: FSH, pituitary; oestrogen, ovaries; LH, pituitary; progesterone, ovaries
Positive interaction of FSH on oestrogen and oestrogen on LH; negative interaction of oestrogen on FSH and progesterone on both FSH and LH
2. a) Peak in LH just before ovulation, just after peak in oestrogen
   b) FSH high between day 1 and before peak in oestrogen; menstruation day 1 to 4; lining build up 4–14; lining maintained 14–28
   c) Lines to show progesterone levels and uterus lining maintained

Main activity: Contraception

1. a) Hormonal: oral contraceptives, implants and patches; some IUDs; the rest are non-hormonal
   b) The methods and descriptions are matched in pairs in the table before being cut out.
   c) Very high reliability: sterilisation, implants, IUDs
   High reliability: oral contraceptive, injections and patches
   Moderate reliability: barrier methods
   Lower reliability: spermicides, periodic abstinence
   d) Advantages and disadvantages: may be based around relative ease of use, reliability as contraceptive, health risks, reversibility, prevention of STIs

Homework activity: Reproduction and hormones quick quiz

1. If the egg is fertilised, it will support the developing embryo.
2. puberty
3. ovulation
4. Approximately every 28 days
5. In the ovary
6. Carried in the bloodstream
7. Causes development of male secondary sexual characteristics at puberty; stimulates sperm production
8. Any two male secondary characteristics from: facial hair, deep voice, pubic hair, broader chest, and so on
9. Female sterilisation
10. progesterone

BIOLOGY HIGHER

23 Homeostasis and response: Treating infertility and negative feedback

Learning objectives

- To explain the use of hormones in modern reproductive technologies to treat infertility
- To explain the roles of thyroxine and adrenaline in the body
- To explain how thyroxine levels are an example of negative feedback

Specification links

- 4.5.3.6
- 4.5.3.7

Starter activity

- **Hormone fast facts; 10 minutes; page 150**

  This activity assesses basic knowledge of the four key hormones for this lesson. Ask the student to fill in the first three bullet points for each activity first and then add further detail if there is time.

Main activities

- **Treating infertility; 15 minutes; page 151**

  Question 1 looks at fertility drugs, the student should answer the questions verbally as a basis for discussion of key points. Question 2 looks at the process of IVF, students should order the statements writing letters in the correct order. Question 3 considers some problems of IVF.

- **Adrenaline, thyroxine and negative feedback; 15 minutes; page 152**

  Question 1 (a verbal activity) considers the effects of adrenaline on the body and how they prepare it for flight or flight. Question 2 covers thyroxine and requires the student to draw a negative feedback loop onto the diagram using the text provided.

- **Social and ethical issues of IVF; 5 minutes**

  Ask the student to describe two social or ethical issues associated with IVF treatments. For example, dealing with 'spare' embryos, cost of treatment (to both individuals and the health service), possibilities for modifying embryos, ability of women with no male partner to give birth to a child using donated sperm and IVF.

Plenary activity

- **Hormones review; 5 minutes**

  First, ask the student to look back at the starter and tell you three details that they would like to add. Then remind them about the negative feedback control of thyroxine and ask the student to think of as many other examples of negative feedback as they can from previous lessons on hormones. This will link to the homework task.

Homework activity

- **Human hormones and negative feedback; 45 minutes; page 153**

  This task asks the student to summarise the topics of homeostasis and hormones, with a focus on negative feedback. Encourage the student to use colour and to develop good practice in summarising notes that will help with revision.

Support ideas

- **Treating infertility** It may help the student to visualise the steps of IVF if they are shown a picture of the embryos that are implanted. This will also emphasise the fact that without modern microscopy techniques IVF would not be possible.
- **Adrenaline, thyroxine and negative feedback** The student may need a reminder of the definition of basal metabolic rate (BMR; see glossary).

Extension ideas

- **Treating infertility** Introduce the idea that other fertility drugs may stimulate the mother's own FSH and LH production. For example, clomiphene blocks the negative feedback action of oestrogen (on FSH and LH production). Ask the student to explain how this works to increase fertility.
- **Adrenaline, thyroxine and negative feedback** Ask the student to produce a flow diagram to show the adrenaline response to danger.

Progress and observations

BIOLOGY HIGHER

Starter activity: Hormone fast facts

Time 10 mins

Learning objectives
- To describe the role of some key hormones

Equipment
none

1. This lesson looks at the roles or uses of four human hormones. Show what you know about these hormones by completing the boxes below to make 'fast facts' cards to compare them.

Follicle stimulating hormone (FSH)	**Luteinising hormone (LH)**
produced by:	produced by:
target cells:	target cells:
effects:	effects:
further information, such as medical uses or conditions:	further information, such as medical uses or conditions:

Thyroxine	**Adrenaline**
produced by:	produced by:
target cells:	target cells:
effects:	effects:
further information, such as medical uses or conditions:	further information, such as medical uses or conditions:

BIOLOGY HIGHER

Main activity: Treating infertility

Time 15 mins

Learning objectives

- To explain the use of hormones in modern reproductive technologies to treat infertility
- To evaluate the methods of treating infertility

Equipment

none

1. Fertility drugs may be given to a woman who finds it difficult to become pregnant. She may then become pregnant in the normal way.

   Discuss with your tutor:
   - What female hormones are present in these fertility drugs?
   - How do they increase the chances of becoming pregnant?

2. In Vitro Fertilisation (IVF) treatment is another infertility treatment that involves the use of hormones. Fertilisation takes place outside of the mother's body, in a laboratory dish – 'in vitro' literally means 'in glass'.

   The table below describes the steps in IVF. They are not in the correct order. Write down the letters in the correct order.

A.	The fertilised eggs are allowed to develop into embryos until the stage when they are tiny balls of cells.
B.	Eggs are fertilised by sperm from the father in the laboratory.
C.	The mother is given FSH and LH.
D.	One or two of the embryos are inserted into the mother's uterus (womb).
E.	The eggs are collected from the mother.
F.	This stimulates several eggs to mature.

   Correct order of IVF steps _____

3. Fertility treatment may give a woman the chance to have a baby of her own but there are problems associated with IVF. For each bullet point below explain to your tutor why IVF carries these problems.
   - The woman may suffer emotionally.
   - The woman may find the process physically stressful.
   - The success rates are not high.
   - It can lead to multiple births, which are a risk to both the babies and the mother.

BIOLOGY HIGHER

Main activity: Adrenaline, thyroxine and negative feedback   **Time** 15 mins

Learning objectives

- To explain the roles of thyroxine and adrenaline in the body
- To explain how thyroxine levels are an example of negative feedback

Equipment

- pencil

Adrenaline is produced by the adrenal glands in times of fear or stress. It prepares the body for physical exertion to deal with harmful situations – often called 'flight or fight'.

Some effects of adrenaline are:

A. It increases the heart rate.

B. It causes blood to be directed away from the digestive system and towards the muscles.

C. It dilates the bronchioles.

D. It causes the liver to convert glycogen to glucose.

1. Explain to your tutor how each of these effects prepares the body for fight or flight.

Thyroxine from the thyroid gland stimulates the basal metabolic rate. It plays an important role in growth and development. Here is a description of how thyroxine production is controlled through negative feedback.

> **Falling levels of thyroxine are detected in the brain (hypothalamus) causing release of a hormone (TRH – thyroid releasing hormone). This acts on the pituitary gland causing it to release thyroid-stimulating hormone (TSH) into the blood. TSH then acts on the thyroid, causing it to increase production of thyroxine. The thyroid gland secretes thyroxine into the bloodstream. When thyroxine levels are high enough, thyroxine in the blood inhibits the release of TRH and TSH. This is negative feedback.**

2. Draw the position of the pituitary and thyroid gland on the diagram. Then add labelled arrows, including positive or negative signs to show the stimulatory or inhibitory effects of hormones in this feedback loop.

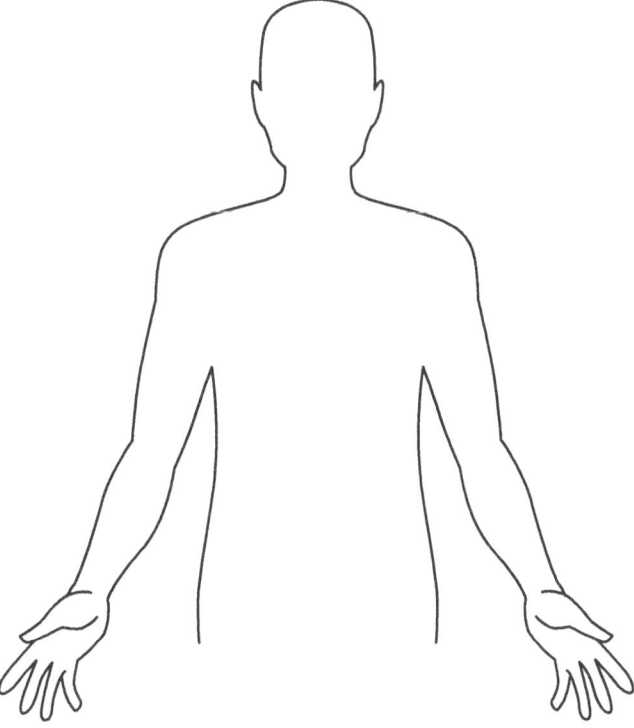

152

BIOLOGY HIGHER

Homework activity: Human hormones and negative feedback   Time 45 mins

Learning objectives
- To describe a range of human endocrine responses
- To explain the role of negative feedback in hormonal control mechanisms

Equipment
- A3 paper

1. Summarise the examples of hormonal control mechanisms that you have learned about. Use the outline below as a starting point, or better still copy it out onto a larger piece paper to give yourself more room. Add diagrams where possible. Many of the examples (all except fight or flight and nitrogen balance) contain elements of negative feedback. Include details about negative feedback on the diagram where appropriate, identifying the stimulus or change in conditions, mechanism of control and response.

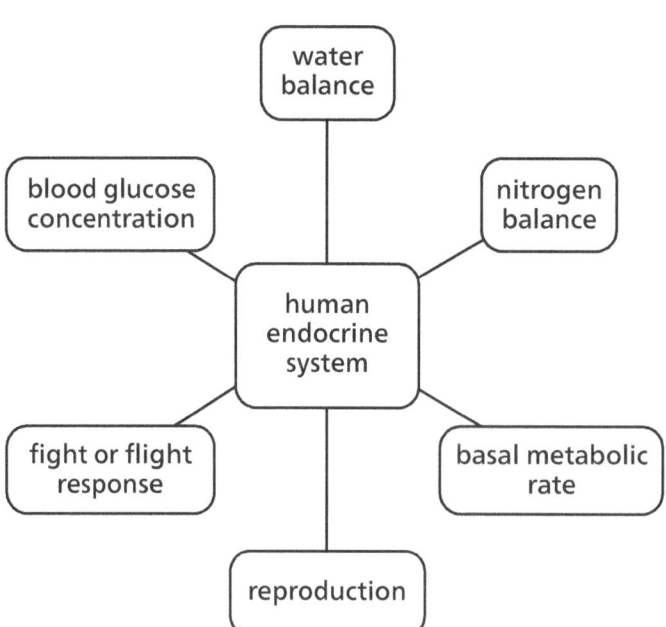

23 Answers

Starter activity: Hormone fast facts

1. FSH: produced by pituitary, targets ovaries, effects causes maturation of an egg
   LH: produced by pituitary, targets ovaries, effects stimulates the release of the egg
   Thyroxine: produced by thyroid gland, target cells in nearly all tissues in the body, effect stimulates the basal metabolic rate and plays an important role in growth and development
   Adrenalin: produced by the adrenal glands, targets cells in numerous tissues such as heart, liver, blood vessels, airways, effect increases heart rate and boosts the delivery of oxygen and glucose to the brain and muscles, preparing the body for 'flight or fight'

Main activity: Treating infertility

1. FSH and LH; they cause egg cells to mature in the ovary and be released; LH can also be used to give precise timing of ovulation
2. C, F, E, B, A, D
3. Emotional: process is stressful as embryos may not implant and will be lost, the treatment may disrupt working and social life, there may be a large financial cost
   Physical: the hormones cause side effects (such as headaches, mood swings) and the procedure may cause discomfort
   Success rates: after several cycles, a woman may still not become pregnant; rates vary from about 32% in women under 35 to about 5% at age 43
   Multiple births: several embryos may be implanted to increase the chance of success, but all may survive; problems include risk of miscarriage, premature births and high blood pressure (pre-eclampsia)

Main activity: Adrenaline, thyroxine and negative feedback

1. A and B provide a greater blood supply to the brain and muscles. This boosts the delivery of oxygen and glucose, which are needed for respiration to release energy for physical exertion or brain activity. C increases air exchange and therefore oxygen. D increases blood glucose levels for respiration in muscle and brain.
2. Minimum labels would be arrows with a positive sign from the brain to the pituitary (labelled TRH) and from the pituitary to the thyroid (labelled TSH), and a negative arrow from the thyroid to the brain/pituitary (labelled thyroxine).

Homework activity: Human hormones and negative feedback

1. The student should record the hormones involved for each example of control. Negative feedback could be shown as a feedback loop for each case. Most will be easily identified. For reproduction, negative feedback includes interactions between hormones. High levels of one hormone may inhibit (prevent) the production of another hormone, for example rising FSH stimulates oestrogen production, which in turn inhibits FSH production, causing levels to fall again.

BIOLOGY HIGHER

24 Homeostasis and response: Plant hormones

Learning objectives

- To explain how plant hormones coordinate and control plant growth responses to light and gravity
- To describe the effects of some plant hormones
- To describe ways people use plant hormones to control growth

Specification links

- 4.5.4.1
- 4.5.4.2

Starter activity

- **Plant responses to light; 5 minutes; page 156**
  This activity should be done independently and will assess the student's basic understanding of phototropism. The student should be able to complete the drawing task easily. Look for appropriate details and use of keywords.

Main activities

- **Phototropism and gravitropism; 15 minutes; page 157**
  Ask the student to complete question 1 independently and check their understanding of key terms. Then work through the investigation examples in questions 2 and 3 together, getting the student to draw on the worksheet.
- **Uses of plant hormones; 15 minutes; page 158**
  Ask the student to complete question 1 and check their knowledge of the roles of each plant hormone. Letters can be written in the table for question 2 to save time. Tissue culture may need to be explained. Questions 3 and 4 consider some uses in more detail and can be done verbally.
- **Role of gravitropism and phototropism; 10 minutes**
  Ask the student to explain the survival advantages to the plant of: positive phototropism in shoots (increased photosynthesis); negative phototropism in roots (roots grow into soil, towards supply of water and ions); positive gravitropism in roots (answer same as two.); negative gravitropism in shoots (plants grow away from soil covering, towards light, more photosynthesis). This can be done verbally.

Plenary activity

- **Gravitropism considerations; 5 minutes**
  Ask the student to look again at the germinating seed in the first main activity. Ask what they would expect to happen if the tips of the shoot and root were removed. Ask them to suggest how they would investigate this.

Homework activity

- **Plant growth investigation; 15 minutes; page 159**
  The homework is an exam-style question focused on required practical activity 8 (investigate the effect of light or gravity on the growth of newly germinated seedlings).

Support ideas

- **Phototropism and gravitropism** It would be helpful to show students houseplants that are showing phototropism, or to set up growing beans or similar in advance to show the geotropism response.
- **Uses of plant hormones** Visual prompts will help a student to remember the uses of plant hormones. For example, demonstrate how rooting powder can be used on a cutting or show internet pictures of explants from tissue culture.

Extension ideas

- **Phototropism and gravitropism** Explain how auxin from shoot tips placed onto agar jelly will enter the jelly. Sketch diagrams to show auxin-impregnated agar blocks placed to the side of or centrally on a shoot; ask the student to predict the effects.
- **Uses of plant hormones** Emphasise the fact that the effect of plant hormones is not always predictable. Different concentrations may have different effects on different tissues; provide examples. Ask students to suggest in general terms how hormones might have an effect.

Progress and observations

BIOLOGY HIGHER

Starter activity: Plant responses to light   **Time** 5 mins

Learning objectives
- To understand the basic plant growth response to light

Equipment
- pencil

1. The diagram shows a newly germinated seedling which has been placed inside a box. Light can only enter from one side.

a) Add to the diagram to show what the plant will look like when it has been left to grow for a few weeks.

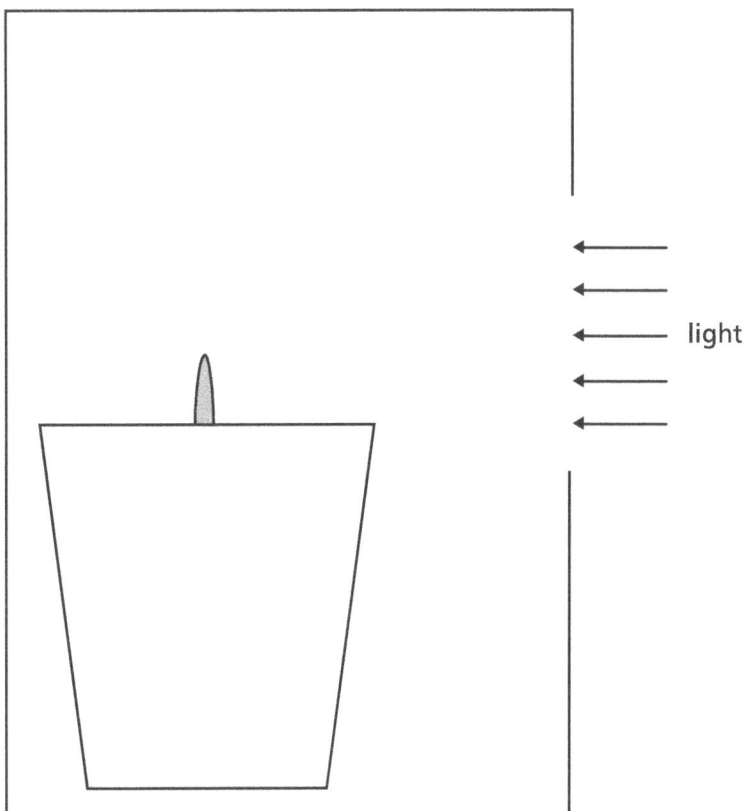

b) Write an explanation for the results in as much detail as you can.

_____
_____
_____
_____

BIOLOGY HIGHER

Main activity: Phototropism and gravitropism

Time 15 mins

Learning objectives
- To explain how plant hormones coordinate and control plant growth responses to light and gravity

Equipment
- pencil
- spare paper

1. **Fill in the gaps to complete the explanation of how plants respond to light. Use the word bank to help you.**

   The growth response of plants to light is called _____ . This response is controlled by the hormone _____ . Auxin is produced in the growing tip and passes down the shoot. Light falling on one side of the shoot causes the auxin to become _____ distributed. There will be more auxin on the shaded side. High auxin concentration causes the cells in the shoot to _____ more. Therefore, the shaded side grows more and the plant bends _____ the light. Growth towards light is called _____ phototropism. In roots, high auxin concentration has the opposite effect, it causes cells to elongate _____ . Plant roots grow away from the light. This is known as _____ phototropism.

   Word Bank: elongate,   positive,   auxin,   phototropism,   negative,   unequally,   towards,   less

2. **There are lots of investigations that can be done to demonstrate the effects of auxin in phototropism. Draw additional lines on each diagram to show how the shoot would grow. Then explain it to your tutor in terms of auxin levels and effects. The arrows show light direction.**

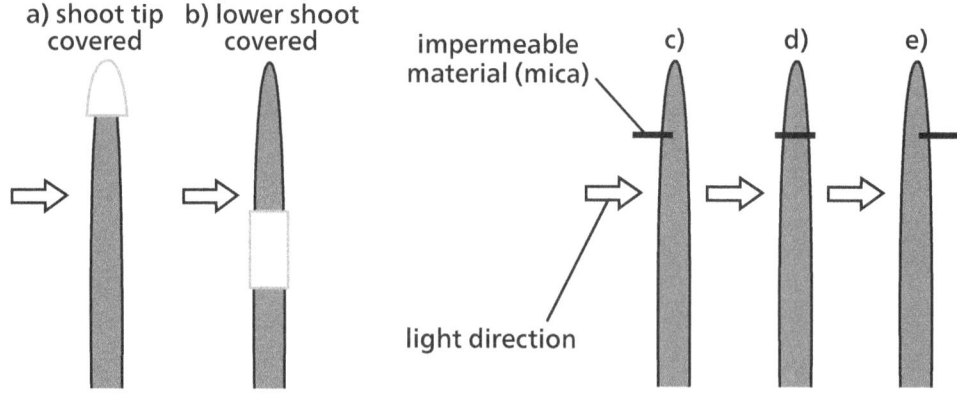

3. **Auxin is also involved in the growth response of plants to gravity. This is called geotropism or gravitropism. The diagram below shows a newly germinated seed with a shoot and root just emerging. Copy the diagram onto a separate piece of paper. Shade in where auxin concentrations would be highest. Then draw on the diagram to show how each would grow. Finally, explain your drawing to your tutor in terms of auxin concentrations and its effects.**

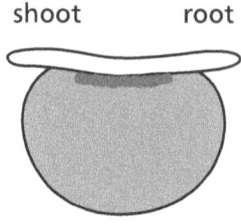

BIOLOGY HIGHER

Main activity: Uses of plant hormones   Time 15 mins

Learning objectives
- To describe the effects of some plant hormones
- To describe ways people use plant hormones to control growth

Equipment
none

1. Draw lines to match the plant hormone to the process it controls in plants.

| auxins | causes seeds to start to germinate and other changes in development |

| gibberellins | controls cell division controls ripening of fruit |

| ethene | controls cell elongation and growth |

2. If you put a ripe banana in a paper bag with unripe fruit it makes it ripen quickly – this is because ripe bananas produce a hormone that controls the ripening process. Plant hormones are useful in many ways in agriculture and horticulture. Look at the list of uses below and complete the table to decide which hormone could be used, based on your knowledge of hormone's roles in plants. Write A, B, C and so on in the correct boxes.

A. to end seed dormancy so that seeds begin to geminate

B. in weedkillers; disrupts growth in broad-leaved plants

C. to increase fruit size, making fruits grow quickly without seeds

D. in rooting powders; used to make roots grow on cuttings of shoots

E. to promote flowering; causes flowers to develop

F. promotes growth in tissue culture (new plants are grown from small pieces of other plants)

G. to control ripening of fruit during storage and transport

Hormone	Auxins	Gibberellins	Ethene
Uses			

3. Plant hormones are used in weedkillers that are sprayed on lawns. Growth in broad-leaved plants, which on lawns are weeds, is more affected that in narrow-leaved grass plants. Why might this be?

_____

_____

4. Plant hormones that control fruit ripening are used in the food industry. Fruit such as tomatoes can be picked when unripe and then ripened when needed using the hormone. Explain to your tutor why this might be helpful to the food industry.

158

BIOLOGY HIGHER

Homework activity: Plant growth investigation   Time 15 mins

Learning objectives

- To explain how the effect of light on newly germinated seedlings can be investigated

Equipment

none

1. Two students wanted to investigate the effect of light direction on the growth of newly germinated mung bean shoots. They compared plants grown with light coming from above with plants grown with light coming from one side. The beans were grown in soil-filled plant pots that were placed inside boxes with openings to the light in different places. Lamps were placed outside the light openings.

    The setup of their investigation is shown in the diagram.

    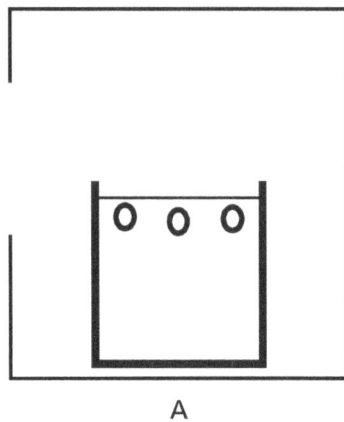
    A
    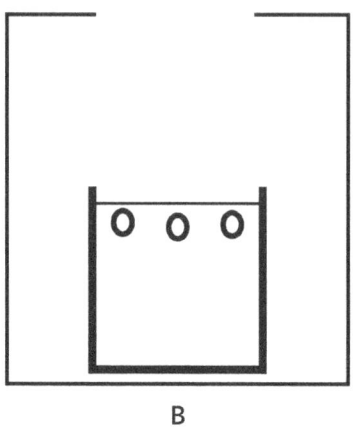
    B

    a) The students wanted to measure the length of the shoots. Suggest one way that they could measure this.

    _____
    _____

    b) Explain why this measurement might be different from the vertical height that the plants grow to.

    _____

    c) The students want to compare the effect of light direction on growth. To do this they need to control other factors that might affect growth. State three factors that they should control and describe how they should control them.

    _____
    _____
    _____
    _____
    _____
    _____
    _____

24 Answers

Starter activity: Plant responses to light

1. a) The plant should be drawn showing growth towards the light.
   b) The explanation should include keywords such as auxin and phototropism. The uneven distribution of auxin and the effect of auxin on cell elongation should be included.

Main activity: Phototropism and gravitropism

1. Phototropism, auxin, unequally, elongate, towards, positive, less, negative
2. a) Shoot does not bend (auxin concentrations released at tip is uniform)
   b) Shoot bends to left (tip produces auxin – is photosensitive, more auxin on shaded side, more cell elongation)
   c) Shoot curves towards light, auxin levels higher away from light
   d) No curved growth as auxin from tip cannot pass down the shoot
   e) No curved growth as auxin from the tip cannot pass down the shaded side of the shoot
3. Auxin concentration is highest at bottom axis of shoot/root in direction of gravity; the shoot grows up (positive geotropism) and the root grows down; high auxin in bottom axis reduces cell elongation in the root but increases cell elongation in shoot

Main activity: Uses of plant hormones

1. Auxins: controls cell elongation and growth; gibberellins: causes seeds to start to germinate; ethene: controls cell division, and ripening of fruit
2. Auxins: B, D, F; gibberellins: A, E, C; ethene: G
3. Broad-leaved plants have a bigger surface area to absorb the weedkiller, so concentrations become higher in the plant.
4. Unripe fruit is firmer and less likely to be damaged during transport. Unripe fruit can be stored in a fridge so they do not rot and then ripened later using the hormone. It extends the season that fruit is available for.

Homework activity: Plant growth investigation

1. a) Allow any method that would take into account the bend in the shoots such as using calibrated string or flexible tape measure; measure from soil to shoot tip.
   b) Shoots in pot A would be bent/grow sideways; the vertical height would be reduced
   c) Examples: light intensity, wavelength or duration; same brightness of lamp, same wavelength from lamp; same length of time each day; stop clock; same amount of water; measure volume or mass with a measuring cylinder/pipette/syringe; mineral ions; same concentration used/same dose; same temperature; use a thermostatically controlled incubator/check temperature is same with a thermometer; size/number of beans; count or suggest number/find mass using balance; any other suitable factor; method

25 Inheritance, variation and evolution: Types of reproduction and meiosis

Learning objectives

- To compare sexual and asexual reproduction
- To explain the process and functions of meiosis and fertilisation
- To describe how some organisms reproduce both sexually and asexually

Specification links

- 4.6.1.1
- 4.6.1.2
- 4.6.1.3

Starter activity

- **Types of reproduction; 5 minutes; page 162**

  This activity explores the student's understanding of the differences between asexual and sexual reproduction.

  They should work independently to place ticks in the asexual and/or sexual columns as appropriate. When complete, review their answers and explain any areas of misunderstanding during the course of the lesson.

Main activities

- **Meiosis; 20 minutes; page 163**

  Ask the student to complete the diagrams one at a time, supporting them as required. Knowledge of the stages of meiosis is not required. Emphasise that the chromosomes occur in matching pairs (the term 'homologous' is not required), that only one of each pair is passed on to each gamete, so that the chromosome number is halved and that allocation is random. Questions 2–4 can be done verbally. In question 5 emphasise that meiosis leads to non-identical cells being formed while mitosis leads to identical cells being formed.

- **Sexual and asexual reproduction; 15 minutes; page 164**

  First, ask the student to explain their understanding of natural selection, a basic understanding is required here. Emphasise that genetic variation is essential for natural or artificial selection. Then ask the student to complete the table, prompting as necessary. For question 2, knowledge of reproduction in organisms is restricted to fungi/strawberry/daffodil and malaria but students should be able to discuss advantages and disadvantages of sexual or asexual reproduction generally.

- **Sexual and asexual reproduction in malaria; 5 minutes**

  Ask the student to describe the basic life cycle of malarial parasites as covered in lesson 12 and to identify where asexual (in the human host) and sexual (in the mosquito) reproduction take place.

Plenary activity

- **Comparing sexual and asexual reproduction; 5 minutes**

  Ask the student to complete a Venn diagram to compare sexual and asexual reproduction. Factors in common include need for cell division and production of new individuals. Differences include type of division and number of parents.

Homework activity

- **Reproduction and cell division; 15 minutes; page 165**

  The homework comprises questions on cell division and asexual/sexual reproduction and should need little explanation.

Support ideas

- **Meiosis** Cell division by meiosis and mitosis can be modelled using pieces of string or modelling material to represent chromosomes.
- **Sexual and asexual reproduction** Show pictures, or preferably real examples of bulbs, flowers, plants with runners, mushroom spore prints (leave a mushroom for a few hours gills down on paper) or pictures from a microscope. This will help the student to visualise the required examples.

Extension ideas

- **Meiosis** Discuss how crossing over causes variation in the gametes so that no two gametes are the same. Ask the student to sketch this using similar diagrams to those in the table.
- **Sexual and asexual reproduction** Ask the student to describe the process of selective breeding using a flow diagram.

Progress and observations

BIOLOGY HIGHER

Starter activity: Types of reproduction Time 5 mins

Learning objectives
- To describe the main differences between sexual and asexual reproduction

Equipment
none

1. For each of the reproductive features or processes described in the table decide whether this applies to asexual or sexual reproduction, or both. Place a tick (✓) in one or both columns as appropriate.

	Reproductive feature or process	Asexual reproduction	Sexual reproduction
1	It involves only one parent.		
2	It involves the joining (fusion) of male and female gametes.		
3	It is involved in reproduction of the malaria parasite.		
4	It involves cell division by mitosis only.		
5	It involves two parents.		
6	Plants may reproduce this way.		
7	It involves cell division by meiosis to produce gametes.		
8	Mixing of genetic information occurs.		
9	The offspring produced show variation.		
10	Involves sperm and egg cells in animals.		
11	Humans reproduce this way.		
12	It involves pollen and egg cells in flowering plants.		
13	Fungi may reproduce this way.		
14	The offspring produced are genetically identical (they are clones).		

BIOLOGY HIGHER

Main activity: Meiosis

Time 20 mins

Learning objectives

- To explain the process and function of meiosis
- To explain how fertilisation leads to development of an embryo

Equipment

- pencil
- spare paper

1. Eukaryotic cells contain pairs of chromosomes. One of each pair came from the mother and one from the father. The diagram below represents a cell with two pairs of chromosomes. Different shading represents chromosomes that were inherited from different parents. Complete the diagram to show what happens when the cell divides by meiosis to form gametes.

(diagram)	a cell with two pairs of chromosomes before meiosis
(diagram)	Copies of the genetic information are made.
	The cell then undergoes the first division of meiosis. One chromosome from each pair is randomly assigned to each of the two daughter cells.
	A second division takes place to form four gametes. Each gamete has a single set of chromosomes. The number of chromosomes in each cell is halved. All the gametes are genetically different from each other. Meiosis is complete.

2. Explain to your tutor where meiosis would take place: a) in mammals b) in plants?

3. In sexual reproduction, explain to your tutor why it is necessary to produce gametes with half the usual chromosome number.

4. After fertilisation, explain to your tutor how the fertilised egg cell (zygote) develops into an embryo and then a new individual?

5. On a separate piece of paper, sketch what would happen if the starting cell in the table above divided by mitosis.

BIOLOGY HIGHER

Main activity: Sexual and asexual reproduction

Time 15 mins

Learning objectives

- To compare the advantages of sexual and asexual reproduction
- To describe how some organisms reproduce using both sexual and asexual methods

Equipment

- bulbs, plants with flowers/seeds and plants with runners
- pencil

1. Complete the sentences in the table to describe the advantages of sexual and asexual reproduction.

Advantages of sexual reproduction	Advantages of asexual reproduction
• there is mixing of genetic information which leads to the offspring showing…	• Only one…
• If the environment changes …	• The organism does not need to find a mate so this saves …
• To increase food production humans can …	• Asexual reproduction is faster than…
	• When conditions are favourable …

2. Some organisms reproduce by both methods depending on the circumstances. Look at the diagrams below and add labels to describe how asexual and sexual reproduction are happening.

| Daffodil | Strawberry |

3. Fungi have complex and varied life cycles which often involve both sexual and asexual stages. Decide with your tutor whether the following examples are sexual or asexual methods.

a) Spores are produced in the parent plant by mitosis. They disperse on the wind and grow into new individual fungi.

b) Some types of fungal cells are produced by meiosis. Pairs of these cells may fuse to make cells that show genetic variation.

Homework activity: Reproduction and cell division

Time 15 mins

Learning objectives

- To compare meiosis and mitosis
- To consider the advantages and disadvantages of asexual reproduction

Equipment
none

1. Complete the table to compare cell division by meiosis and mitosis. Some boxes have been completed for you.

	Mitosis	Meiosis
Number of divisions		
Number of daughter cells produced		
Number of chromosomes in daughter cells	same as the original cell	
Genetic make-up of daughter cells	identical to each other and identical to the original cell	
Role in reproduction		

2. A market gardener has developed an attractive variety of garden plant by selective breeding. She wants to reproduce from this plant so that the attractive variety can be marketed. She decides to use cuttings to produce her new plants. Taking cuttings is an artificial method of asexual reproduction. Small pieces of the plant are removed and stimulated to grow into whole new plants.

   Suggest one advantage and one disadvantage of using asexual reproduction to produce new plants.

   _____
   _____
   _____
   _____
   _____

25 Answers

Starter activity: Types of reproduction
1. 1 A; 2 S; 3 A, S; 4 A; 5 S; 6 A, S; 7 S; 8 S; 9 S; 10 S; 11 S; 12 S; 13 A, S; 14 A

Main activity: Meiosis
1. Student's own answers. Allow an alternative assortment, such as two of same colour in each cell.

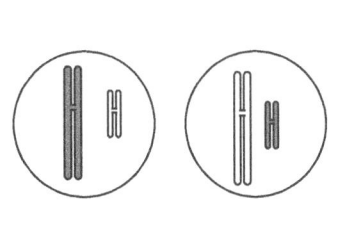 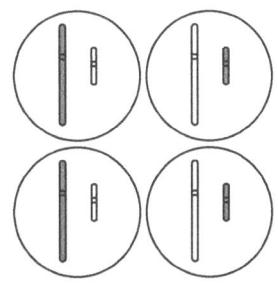

2. a) Testes and ovaries
   b) Anthers and ovaries of flowers
3. So that the normal number of chromosomes is restored after fertilisation
4. The new cell (zygote) divides by mitosis, so the number of cells increases. As the embryo develops the new cells differentiate.
5. Steps one and two would be the same as in the table. Two daughter cells identical to the initial cell would be produced in step three.

Main activity: Sexual and asexual reproduction
1. Sexual: variation/variation gives a survival advantage by natural selection/speed up natural selection by selective breeding
   Asexual: parent needed/time and energy/sexual reproduction/lots of identical offspring can be produced to exploit resources quickly
2. Daffodil: sexually using flowers and seeds, asexually by bulb division; strawberry: reproduces sexually by flowers and seeds (on surface of fruit), asexually by runners
3. a) asexual
   b) sexual

Homework activity: Reproduction and cell division
1.

	Mitosis	Meiosis
Number of divisions	1	2
Number of daughter cells	2	4
Number of chromosomes	Same as the original cell	Half the number of the original cell
Genetic make-up of daughter cells	Identical to each other and to the original cell	Different from each other and different from the original cell
Role in reproduction	Asexual reproduction	Produces gametes for sexual reproduction

2. Advantages: only one parent plant is needed so the best plant can be used; faster than sexual reproduction/growing from seeds; will produce genetically identical offspring so all will show the attractive features/in sexual reproduction the offspring would show variation and may not have the attractive features
Disadvantages: there will be no variation in offspring so unable to adapt to changes environment/if parent is susceptible to environmental change or disease then all offspring will be; no variation so no possibility for future selective breeding

BIOLOGY HIGHER

26 Inheritance, variation and evolution: DNA structure and the genome

Learning objectives

- To describe the structure of DNA
- To discuss the importance of understanding the human genome
- To describe protein synthesis and the effects of mutations

Specification links

- 4.6.1.4
- 4.6.1.5

Starter activity

- **DNA and genes; 5 minutes; page 168**

  The student should work independently to show their existing knowledge of DNA structure and to define a gene. Do not correct the student's answer to question 1 as this will be revisited in the main activities and in the plenary. Make sure that the student's understanding of the term 'gene' is clear.

Main activities

- **DNA structure and the genome; 15 minutes; page 169**

  Discuss question 1 verbally initially. Emphasise that the genome is not the same as the genes that a person has. Work through question 2, one step at a time, encourage the student to work independently using the text in italics to help. Questions 3–5 can be discussed verbally. For question 4, ensure the student understands that only one of the strands contains the code.

- **Protein synthesis and mutations; 15 minutes; page 170**

  The student is not expected to know or understand the structure of mRNA, tRNA, or the detailed structure of amino acids or proteins. Questions can be answered verbally apart from 1 and 2 which should be answered on the sheet.

- **The importance of understanding the human genome; 10 minutes**

  Ask the student to suggest why developments in understanding the whole human genome are important and useful. They should record their ideas in a bullet point list or spider diagram. Key ideas are that it is likely to become more important for medicine in the future, it allows research to identify the genes linked to different types of disease (such as cancer), it helps understanding and treatment of inherited disorders and is also used to trace human migration patterns.

Plenary activity

- **DNA structure revisited; 5 minutes**

  Ask the student to look again at their diagram from the starter. Using a different colour pen they should amend this diagram and add detail. They may wish to draw an additional diagram. Discuss their learning and make sure that the following key ideas are secure: DNA is made up of two strands in a double helix; sugar, phosphate, bases A, C, G and T; that C links to G and T to A on complementary strands; that three bases code for one amino acid.

Homework activity

- **The human genome and protein synthesis; 15 minutes; page 171**

  These short, exam-style questions review the learning in the main activities.

Support idea

- **DNA structure and the genome** Show animations of DNA to reinforce how the model relates to the double helix structure. A web search for 'DNA double helix animation' will give several useful results.

Extension ideas

- **DNA structure and the genome** Ask the student to suggest how many different amino acids could be coded for with four bases in triplets. The answer is 64 (4 × 4 × 4), more than enough to code for the 20 occurring amino acids. Discuss this redundancy.
- **Protein synthesis and mutations** Show a table linking DNA base triplet code to specific amino acids (a search for DNA triplet code will give suitable results). Give example triplets and ask the student to find the amino acid linked to that triplet. Discuss the effect of a single base substitution.

Progress and observations

Starter activity: DNA and genes

Time 5 mins

Learning objectives
- To describe the structure of DNA
- To understand the relationship between genes, DNA and proteins

Equipment
- pencil

1. In the space below, draw a diagram to show the structure of DNA. Add labels to show as much detail as you can.

2. What is a gene? Write your own definition.

_____
_____
_____

BIOLOGY HIGHER

Main activity: DNA structure and the genome

Time 15 mins

Learning objectives
- To describe the structure of DNA
- To define the term genome

Equipment
- blue and yellow pencils or felt tip pens

1. **Explain to your tutor what the term 'genome' means. Then write a definition below.**

   _____

   _____

DNA is a polymer made up of repeating nucleotide units. Each nucleotide consists of a common sugar and phosphate group with one of four different bases attached to the sugar. There are four different nucleotides depending on which base is present, either A, C, G or T. The nucleotides are joined together into long strands of DNA consisting of alternating sugar and phosphate sections. DNA is made up of two strands of nucleotides in a double helix shape.

2. **Using the description to help you, make the following changes to the diagram of DNA:**

   a) On one strand, colour the sugar groups blue and the phosphate groups yellow.

   b) Circle and label any single nucleotide.

   c) Circle an example of each of the other three types of nucleotide.

   d) Write in the letters of the bases on the unlabelled complementary strand.

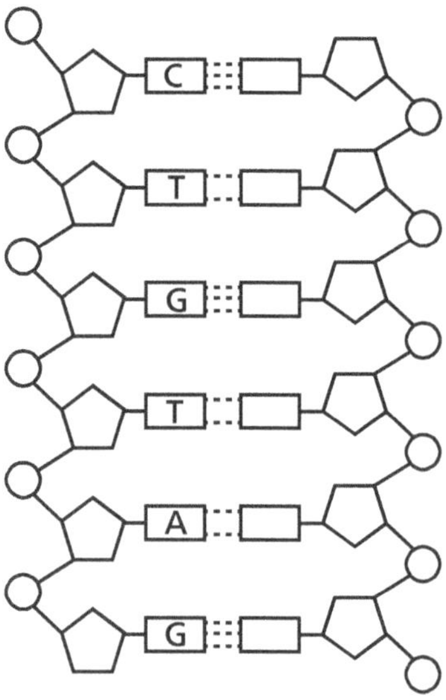

3. **The double helix shape is not shown in the diagram. Explain to your tutor what is meant by the term 'double helix'.**

4. **A sequence of three bases codes for a particular amino acid. How many amino acids would be coded for by the DNA strand in the diagram?**

   _____

5. **What structures within the nucleus contain the DNA?**

   _____

   _____

169

BIOLOGY HIGHER

Main activity: Protein synthesis and mutations   Time 15 mins

Learning objectives
- To give a simple description of protein synthesis
- To explain how the structure of DNA, including mutations, affects the protein made and phenotype

Equipment
- pencil
- spare paper

1. Use the word bank to complete the gaps in these sentences, to describe how proteins are made (synthesised).

   When proteins are synthesised, the base sequence code from the _____ is first transferred to a template molecule. This template of the base sequence is called mRNA. The template moves out of the _____ and attaches to a small structure in the cell called a _____ . Carrier molecules (tRNA) bring specific amino acids to add to the growing protein _____ , one at a time, in the correct order. Each sequence of _____ bases codes for a particular amino acid. The order of the groups of three bases controls the order in which _____ are joined together to produce a particular protein. When the protein chain is complete it _____ to form a unique three-dimensional shape. This unique shape enables the proteins to do their job. Examples of proteins include _____ , hormones, or proteins that form structures in the body such as _____ . There are lots of ribosomes in a cell, so lots of protein molecules can be made at one time.

   | Word bank: | chain | nucleus | enzymes | three | amino | acids | collagen | folds up | DNA | ribosome |

2. Now, on a separate piece of paper, draw a simple diagram to show the process described.

3. Explain the term 'mutation' to your tutor.

   Mutations in an organism's DNA occur frequently. Most do not alter the protein, or only alter it slightly so that its function is not changed. However, sometimes mutations code for an altered protein.

4. Reorganise the statements below and use them to complete the flow diagram to show how a mutation leads to a change in a protein.

   1. One or more bases are deleted, substituted or added to the DNA of a gene.

   - There is a change in the base sequence of the template molecule.
   - There is a change in the amino acid sequence at the ribosome.
   - The final protein has a different shape when it folds.
   - There is a change in the base sequence of DNA.

5. Mutations in DNA can affect the activity of a protein. Explain to your tutor why this might be the case for a) an enzyme and b) a structural protein.

6. Not all parts of DNA code for proteins. It is estimated that more than 98% of your DNA does not code for a protein. Mutations in non-coding DNA can still affect the phenotype of an organism. Explain to your tutor why this is.

BIOLOGY HIGHER

Homework activity: The human genome and protein synthesis Time 15 mins

Learning objectives
- To discuss the importance of understanding the human genome
- To understand a simple description of protein synthesis

Equipment
none

1. The Human Genome Project studied the whole sequence of human DNA.

a) Explain how this knowledge may be used for medicine in the future.

_____

_____

_____

b) Describe one other way in which knowledge of the human genome can be used.

_____

_____

2. The following molecules are involved in protein synthesis:

A. template molecule (mRNA)

B. DNA

C. carrier molecule (tRNA)

D. amino acids

Write letters in the table to match each molecule to its function in protein synthesis.

Function	Letter
are joined together in the correct order to make a particular protein	
carries a copy of the code for a protein from the nucleus to the ribosome	
brings a specific amino acid from the cytoplasm to the ribosome	
contains a sequence of bases, sections of which may code for proteins or may be non-coding	

26 Answers

Starter activity: DNA and genes

1. Do not correct the student response. This will be revisited later in the lesson.
2. A gene is a small section of DNA on a chromosome that codes for a specific sequence of amino acids. This sequence makes a particular protein.

Main activity: DNA structure and the genome

1. The genome of an organism is the entire genetic material of that organism including the DNA that codes for proteins (the genes) and the non-coding parts of the DNA.
2. a) Pentagons of one strand in blue, circles yellow
   b) and c) T, A, C and G nucleotide circled (base along with pentagon sugar and circle phosphate group)
   d) C linked to G and T to A so that reading down the strand, the complementary strand is: G, A, C, A, T, C
3. Double spiral
4. two
5. chromosomes

Main activity: Protein synthesis and mutations

1. DNA, nucleus, ribosome, chain, three, amino acids, folds up, enzymes, collagen
2. Student's own answer. Ensure their diagram shows the different stages of protein synthesis described in part a).
3. A mutation is a change in the sequence of bases in DNA. For example, bases can be added, deleted or substituted.
4. 2: Change in the base sequence of DNA; 3: change in the base sequence of the template molecule (mRNA); 4: change in the amino acid sequence at the ribosome; 5: the final protein has a different shape when it folds
5. a) Shape of the active site of the enzyme may no longer fit the substrate binding site
   b) May lose its strength
6. Non-coding parts of DNA can switch genes on and off, so variations in these areas of DNA may affect how genes are expressed, as genes may not be switched on or off effectively.

Homework activity: The human genome and protein synthesis

1. a) Can search for genes linked to different types of disease; helps doctors to understand and treat inherited disorders
   b) Can be used in tracing human migration patterns from the past

2.

Function	Letter
Are joined together in the correct order to make a particular protein	D (amino acids)
Carries a copy of the code for a protein from the nucleus to the ribosome	A (template molecule – mRNA)
Brings a specific amino acid from in the cytoplasm to the ribosome	C (carrier molecule – tRNA)
Contains a sequence of bases, sections of which may code for proteins or may be non-coding	B (DNA)

BIOLOGY HIGHER

27 Inheritance, variation and evolution: Inheritance and sex determination

Learning objectives

- To be able to interpret genetic diagrams and construct Punnett diagrams to show sex inheritance and monohybrid crosses
- To understand how some disorders are inherited
- To evaluate the economic, social and ethical issues of embryo screening

Specification links

- 4.6.1.6
- 4.6.1.7
- 4.6.1.8
- MS 1c, 2e

Starter activity

- **Genetic definitions; 5 minutes; page 174**

  Provide the student with a cut out set of the boxes. Ask the student to match the terms to the relevant definition.

Main activities

- **Genetic crosses and sex determination; 15 minutes; page 175**

  Ask the questions verbally, except for questions involving diagrams. Ensure that the genetic basis of sex is understood. Emphasise that higher students may be expected to construct their own Punnett diagrams and that the ratios given are theoretical and may not be observed with small numbers of offspring because each event of the joining of gametes is random. The second question looks at fur colour inheritance in mice as an example of a feature controlled by a single gene with multiple alleles.

- **Family trees and genetic disorders; 15 minutes; page 176**

  The activities cover inherited characteristics of red-green colour blindness and polydactyly. Knowledge of cystic fibrosis as a recessive disorder is also required, so it should be mentioned if time permits. It is also covered in the homework. Questions 1 a) and 2 a) can be done verbally; the others should be answered on the sheet.

- **Issues of embryo screening; 10 minutes**

  Students should be able to make informed judgements about the economic, social and ethical issues surrounding embryo screening. Discuss the idea that embryo screening may reduce human suffering and consider the ethical issues which may arise. Ask the student to record the ethical, economic and social arguments for and against screening of embryos in a table during the discussion. There is some overlap here with lesson 23 (IVF).

Plenary activity

- **Note to tutor; 5 minutes**

  Ask the student to write down the most important thing they learned and one question they still wish to ask.

Homework activity

- **Genetic disorders and genetic screening; 30 minutes; page 177**

  The homework covers longer exam-style questions on cystic fibrosis and the ethics of screening embryos.

Support ideas

- **Genetic crosses and sex determination** Take two coins and designate heads and tails as two different alleles, such as brown B/blue b eyes (so each coin is heterozygous). Ask the student what the expected ratio of brown:blue eyed phenotype would be in a family with four offspring. Compare with the outcome of four tosses of the pairs of coins. Repeat to show the effect of chance events on offspring ratios.
- **Family trees and genetic disorders** If the student struggles with the family tree, start with more simple questions. Ask them to point to an unaffected male and female and ask what the various vertical and horizontal lines might represent.

Extension ideas

- **Genetic crosses and sex determination** Introduce the idea of test crosses to determine unknown genotypes.
- **Family trees and genetic disorders** Ask the student to explain why colour blindness is more common in males than females.

Progress and observations

Starter activity: Genetic definitions

Time 5 mins

Learning objectives
- To be able to explain genetic terms

Equipment
- cut out set of the boxes below to make sort cards

1. Your tutor will give you a set of cards that are jumbled up. Match the definitions to the genetics terms.

phenotype	the physical (or chemical) characteristics of an organism, influenced by expression of genes
genotype	the alleles that an organism has
heterozygous	when the two alleles present are different for a particular gene
homozygous	when the two alleles present are the same for a particular gene
dominant	an allele that is always expressed even if only one copy is present; it overrides a recessive allele
recessive	an allele that is only expressed if two copies are present and no dominant allele is present
gene	small section of a chromosome that carries the information for a particular characteristic
alleles	alternative forms of a particular gene
chromosomes	structures found in the nucleus that carry genetic information
gamete	name for a male or female sex cell

BIOLOGY HIGHER

Main activity: Genetic crosses and sex determination

Time: 15 mins

Learning objectives
- To be able to carry out a genetic cross to show sex inheritance
- To understand that most characteristics are a result of multiple genes interacting, rather than a single gene
- To construct a monohybrid genetic cross using a Punnett diagram

Equipment
none

1. **Ordinary human body cells contain 23 pairs of chromosomes. Of these, 22 pairs control characteristics only, but the 23rd pair carries the genes that determine sex. Therefore, it is the inheritance of a pair of particular chromosomes that determines gender.**

    a) Describe the difference between the chromosomes of males and females in humans.

    b) Complete the genetic cross and Punnett square diagram to show the probability of having a boy or a girl.

    Parental phenotype   Male   Female

    Parental genotype   _____   _____

    Gametes   ◯ or ◯   ◯ or ◯

    Punnett square:

    Ratio of male to female offspring _____ : _____

    Proportion of offspring expected to be female _____

    c) Why do the numbers of male and female children in a family not always follow this expected pattern?

2. **Most inherited characteristics are a result of multiple genes interacting, rather than a single gene.**

    a) Describe some examples of characteristics that are influenced by the interaction of multiple genes.

    b) On one gene in mice the allele for black fur (B) is dominant to the allele for brown fur (Bb).

    i) What genotypes are possible for a mouse that has black fur?

    ii) Complete the genetic diagram to show what fur colours you would predict in offspring produced by two mice that are Bb.

    Parental phenotype   _____   _____

    Parental genotype   _____   _____

    Gametes   ◯ or ◯   ◯ or ◯

    Punnett square – genetic cross:

    Predicted ratios of phenotypes in offspring _____ : _____

    Proportion of the offspring expected to be brown _____

175

BIOLOGY HIGHER

Main activity: Family trees and genetic disorders   Time **15** mins

Learning objectives

- To extract and interpret information from family trees
- To explain how some disorders are inherited using genetic crosses

Equipment

- spare paper
- pencil

1. One characteristic that is controlled by a single gene in humans is red-green colour blindness. The gene for red-green vision is sex-linked. It is only found on the X chromosome – there is no matching gene on the Y chromosome. On the X chromosome, the normal red-green vision allele XB is dominant to colour blindness Xb. The Y chromosome has no corresponding allele.

   a) Write down the genotype of a colour-blind female: _____ and a colour-blind male: _____

   b) Family trees can be used to show the inheritance of genetic traits over different generations. An example for red-green colour blindness is given below.

   i) Name one person who we can be certain does not carry the colour-blind allele. _____

   ii) Name two people who we can be certain do carry the colour-blind allele. _____

   iii) What is the probability that Jo is a carrier for the colour-blind allele? _____

   iv) What is Rob's genotype for this trait? _____

2. Polydactyly is a genetic disorder that is caused by the inheritance of a dominant allele.

   a) What are the symptoms of polydactyly? _____

   b) On a separate piece of paper draw a genetic diagram to show what phenotypes you would predict in the children when one parent is heterozygous for polydactyly and one parent does not have the condition. Use D to represent the dominant allele and d to represent the unaffected allele.

176

BIOLOGY HIGHER

Homework activity: Genetic disorders and genetic screening   Time 30 mins

Learning objectives

- To evaluate the economic, social and ethical issues concerning embryo screening

Equipment

none

1. **Cystic fibrosis is a genetic condition caused by a recessive allele.**

   a) What are the symptoms of cystic fibrosis?

   b) What is the probability of a child of two parents that are heterozygous for the cystic fibrosis allele suffering from cystic fibrosis?

2. Answer this exam-style question.

   > If two parents are carriers for a recessive genetic condition, embryo screening can be used to ensure that only healthy babies are born. This involves producing several embryos using IVF, screening the embryos to check for faulty genes and choosing only healthy embryos to implant in the mother's womb.
   >
   > Using your knowledge of IVF and embryo screening, evaluate the ethical, social and economic arguments for and against the use of embryo screening for inherited genetic disorders.
   >
   > [6 marks]

BIOLOGY HIGHER

27 Answers

Starter activity: Genetic definitions

1. Answers on the intact starter activity sheet are correct.

Main activity: Genetic crosses and sex determination

1. a) In females the sex chromosomes are the same (XX). In males the chromosomes are different (XY).
   b) Genotype XY, XX; gametes X or Y and X or X
   Ratio 1 female : 1 male
   Proportion ½/0.5 (or 50%)
   c) Because it is due to chance whether the sperm that joins with an egg carries a Y or an X chromosome; each event is random and not influenced by the outcome for any earlier offspring

	X	X
X	XX	XX
Y	XY	XY

2. a) Examples might include height, weight, skin colour, eye colour
   b) i) BB or Bb
   ii) Phenotype black, black; genotype Bb, Bb; gametes B or b and B or b; ratio 3 black : 1 brown; proportion ¼/0.25 (or 25%)

	B	b
B	BB	Bb
b	Bb	bb

Main activity: Family trees and genetic disorders

1. a) Xb Xb, Xb Y
   b) i) Arthur, Andy or Chris
   ii) Anne and Jan
   iii) 0.5
   iv) XbY
2. a) Having extra fingers or toes
   b) Genotype Dd, dd; gametes D or d and d or d; ratio 1 having polydactyly : 1 unaffected; proportion ½/0.5 (or 50%)

	d	d
D	Dd	Dd
d	dd	dd

Homework activity: Genetic disorders and genetic screening

1. a) Examples: thick mucus builds up in lungs; frequent chest infections; digestive problems
   b) 0.25 (allow 25%)
2. The following table provides guidance on what a Level 3, 2 or 1 answer to this question would look like and the number of marks each would attract.

L3	A detailed, balanced and well organised consideration of arguments for and against. Evaluates arguments, makes a justified conclusion balancing benefits against costs.	5–6 marks
L2	An attempt to evaluate arguments is made. Arguments may have weaknesses in balance, organisation or any justified conclusion.	3–4 marks
L1	Some relevant ethical, social and economic arguments are made.	1–2 marks

Indicative content *Ethical, social or economic arguments for:*	**Indicative content** *Ethical, social or economic arguments against:*
• Prevents having child with the disorder/prevent future suffering/reduce incidence of the disease	• 'Spare' embryos, need to be dealt with, some people may consider destroying them to be murder
• Spare embryo cells could be used in stem cell treatment	• Cost of treatment (to individuals or health service)
• Reduces long-term cost of treating children with the disorder	• Health risk to mother from IVF
• May be ethically more acceptable than abortion of babies tested during pregnancy	• May lead to damage to embryo/embryos cannot give consent
• Gives time for parents to become prepared	• Qualified religious argument
	• Screened embryos may be miscarried/relatively low success IVF
	• May be seen as/may increase prejudice against disabled people
	• May eventually lead to producing designer babies

178

BIOLOGY HIGHER

28 Inheritance, variation and evolution: Genetic understanding, variation and selective breeding

Learning objectives	Specification links
• To describe how the genome and the environment influence phenotype	• 4.6.2.1
• To describe the development of our understanding of genetics	• 4.6.2.3
• To explain the impact of selective breeding of food plants and domesticated animals	• 4.6.3.3
	• MS 2c

Starter activity

- **Causes of variation; 5 minutes; page 180**

  The student should complete the questions independently, then you should review their answers.

Main activities

- **Understanding inheritance; 15 minutes; page 181**

  Ask the student to read the first task and suggest the order of the discoveries. Once this is correct, they should draw lines. Items in bold are specification requirements. Emphasise that the examples are just key points; research on inheritance involved many more scientists.

- **Selective breeding; 15 minutes; page 182**

  Ask the student to complete questions 1 and 2 independently. Discuss their answers and emphasise that only variation determined by genes can be selectively bred. Questions 4 and 5 can be done verbally if time is short.

- **Mutations and variation; 10 minutes**

  Remind the student that there is usually extensive genetic variation within a population of a species and ask how the different genetic variants have arisen (mutations). Review the student's understanding of mutations. Ask the student what would happen if a new phenotype is suited to an environmental change. Ask the student to write three bullet points to summarise how this can lead to a relatively rapid change in the species.

Plenary activity

- **One minute on the history of understanding inheritance; 5 minutes**

  Ask the student to attempt to talk for one minute on the subject of the history of understanding inheritance, without repetition, hesitation or deviation. They might mention understanding of inheritance shown thousands of years ago through selective breeding, as well as research on genes and DNA during the last two centuries.

Homework activity

- **Changes in phenotype; 15 minutes; page 183**

  There are questions about mutations and variation which includes drawing a graph.

Support ideas

- **Understanding inheritance** Show the developments in understanding as a timeline. An internet search for 'history of genetics timeline' will give several useful images.
- **Selective breeding** Show the student a picture of a wild boar and large white pig, ask them to suggest how natural selection may have led to today's farmed pigs.

Extension ideas

- **Understanding inheritance** Discuss the background of the key scientists in more detail, especially Mendel. An internet search for 'DNA from the beginning' will give some useful biographies.
- **Selective breeding** Ask the student to suggest why a study of the genes of wild bananas is important in the fight against Panama disease. Emphasise the genetic variation of the wild population compared with the problems caused by lower genetic diversity of cultivated varieties.

Progress and observations

Starter activity: Causes of variation

Time 5 mins

Learning objectives

- To describe how the genome and the environment interact to influence the development of the phenotype of an organism

Equipment

none

1. Variation is another name for the differences between individuals in a population. Variation may be due to differences in the genes they have inherited (genetic causes) or the conditions in which they have developed (environmental causes) or a combination of both. For the human characteristics in the table, place a tick (✓) in one or both columns to show what causes variation in that characteristic.

Characteristic	Variation caused by genes	Variation caused by environment
height		
eye colour		
weight		
ability to roll tongue		
which language is spoken		
hair colour		
gender		
scars on skin		
skin colour		

2. How does the variation in characteristics caused by genes arise?

   _____

3. How frequently do you think that these new phenotype variants arise?

   _____
   _____
   _____

BIOLOGY HIGHER

Main activity: Understanding inheritance　　　　　　　　　　**Time** 15 mins

Learning objectives

- To describe the development of our understanding of genetics
- To understand why the importance of Mendel's discovery was not recognised until after his death

Equipment

none

1. Our current understanding of genetics has developed over time. Work by many scientists led to the gene theory of inheritance that you learn about today. Some of the major advances in understanding are listed below. Decide what order these occurred in, then draw lines to link to their date of discovery.

Mid-19th century	**The structure of DNA was determined and the mechanism of gene function worked out.** Oswald Avery identified DNA as the substance responsible for heredity. Research by Watson and Crick, and Franklin and Wilkins showed the double helix structure of DNA and how the sugar, phosphate and bases are arranged.
Late 19th century	It was observed that chromosomes and Mendel's 'units' behaved in similar ways. This led to the idea that the 'units', now called genes, were located on chromosomes. Three European scientists independently rediscovered and confirmed Mendel's results, which were now easier to understand because of the observations of chromosome behaviour that had been made.
Early 20th century	Gregor Mendel, an Austrian monk, carried out breeding experiments on plants, especially peas. One of his observations was that the **inheritance of each characteristic is determined by 'units' that are passed on to descendants unchanged.** He also put forward the idea of dominant traits.
Mid-20th century	**Behaviour of chromosomes during cell division was observed.** Walther Flemming, working at the University of Kiel in Germany, first used stains that allowed him to see chromosomes under a microscope during mitosis. He was able to see that chromosomes 'doubled' before dividing into daughter cells.

2. The importance of Mendel's discovery was not recognised until after his death – why was this? Discuss each of the possible reasons with your tutor and add a sentence to each to provide explanation.

   a) He was a monk, not a professional scientist. _____

   b) His main publication 'Experiments in plants hybridisation' was published in an obscure local natural history society journal.
   _____

   c) The accepted theory of inheritance at the time was of 'mixing, or blending' of parents' characteristics in offspring. Mendel's work showed that individual characteristics were inherited unchanged.
   _____

   d) Scientists didn't understand how important his work was. _____

181

Main activity: Selective breeding   Time 15 mins

Learning objectives
- To explain the impact of selective breeding of food plants and domesticated animals

Equipment
none

1. Selective breeding is used by humans to develop plants and domesticated animals with desirable characteristics. Read the statements below and circle either true or false next to each one.

Selective breeding has been carried out only since humans have understood genes.	T F
Today's crop plants were developed by selective breeding from wild plants.	T F
Selective breeding relies on asexual reproduction.	T F
Selective breeding is also known as artificial selection.	T F
Any type of variation can be used in selective breeding.	T F

2. The statements in the boxes describe the process of selective breeding. Decide what order they should go in and write down the letters to show this.

A. Parents with the desired characteristic are chosen.	B. These offspring are bred together.	C. The chosen parents are bred together.
D. Offspring are chosen which show the desired characteristic.	E. The initial population shows variation with a mixture of genetic characteristics.	F. This is repeated over many generations until all the offspring show the desired characteristic.

Answer _____

3. In selective breeding, characteristics are chosen for their usefulness or appearance. For each of the examples below, suggest one feature that has been selected for from the original wild plant or animal.

   a) food crops such as wheat _____

   b) farm animals such as cows _____

   c) domestic dogs _____

   d) flowers _____

4. Describe to your tutor some possible problems of selective breeding.

5. Wild daffodils have small flowers with narrow petals. Explain how selective breeding has produced daffodils with large flowers that are grown in gardens today.

   _____
   _____
   _____

BIOLOGY HIGHER

Homework activity: Changes in phenotype

Time 15 mins

Learning objectives
- To understand the link between mutations and genetic variation
- To be able to draw a graph to show variation

Equipment
- pencil

1. **Fill in the gaps in the sentences below. Use the words from the word bank to help you.**

   There is usually extensive genetic _____ within a population of a species. Variation may also be caused by the _____ that an animal or plant lives in. All genetic variants arise from _____ in DNA. Most of these have no effect on the _____ . Some interact with other genes to influence the phenotype. Only very rarely will a mutation will lead to a _____ phenotype. Mutations occur _____ . If a new phenotype is well suited to an environmental change, it can lead to a relatively rapid change in the _____ .

   Word bank: environment, variation, mutations, phenotype, new, species, continuously

2. **The table shows the results of an investigation into the variation in eye colour in one year 11 class. Decide what sort of graph would be best to use, then draw the graph on the grid below.**

Eye colour	Number of students
blue	18
brown	8
hazel or green	12

28 Answers

Starter activity: Causes of variation

1.

Characteristic	Variation caused by genes	Variation caused by environment
Height	✓	✓
Eye colour	✓	
Weight	✓	✓
Ability to roll tongue	✓	
Which language is spoken		✓
Hair colour	✓	✓
Gender	✓	
Scars on skin		✓
Skin colour	✓	✓

2. mutations
3. While mutations happen continuously, those that change phenotype happen very rarely.

Main activity: Understanding inheritance

1. Mid-19th Mendel; late 19th chromosome behaviour; early 20th chromosomes as Mendel's 'units'; mid-20th structure of DNA and gene function
2. a) His work may not have been taken seriously/it was less likely to become well known in the scientific world
   b) The publication had only a small distribution (only a couple of hundred people). As a local rather than an international publication, it was less likely to be translated into other languages.
   c) Mendel's work contradicted thinking at the time and so was less likely to be accepted.
   d) Mendel's work was not yet supported by other research and understanding. The mechanisms for inheritance of 'units' were not understood.

Main activity: Selective breeding

1. F, T, F, T, F
2. E, A, C, D, B, F
3. Answers could include:
   a) Disease resistance
   b) Increased meat or milk production
   c) A gentle nature
   d) Large or unusual flowers
4. Can lead to 'inbreeding'; the lack of genetic variation may mean that some breeds are particularly prone to disease; selection of some features in pedigree animals can lead to inherited defects; there may be ethical concerns about animal suffering
5. Wild daffodils with largest flowers chosen and cross pollinated, seeds grown and offspring selected with largest flowers, these are bred together, repeat until all offspring have large flowers

Homework activity: Changes in phenotype

1. Variation, environment, mutations, phenotype, new, continuously, species
2. Bar chart chosen, correct scales chosen for axes (must maximise the area of the grid used), vertical axis and bars correctly labelled, all bars correctly plotted, allow plotted points if not a bar chart

184

BIOLOGY HIGHER

29 Inheritance, variation and evolution: Genetic engineering and cloning

Learning objectives

- To describe the process of genetic engineering
- To describe the potential benefits and risks of genetic engineering
- To describe some methods of cloning in plants and animals

Specification links

- 4.6.2.4
- 4.6.2.5

Starter activity

- **GM and cloning crossword; 5 minutes; page 186**

  Ask the student to complete the crossword independently, then go through the answers and check understanding of these key terms.

Main activities

- **Genetic engineering; 15 minutes; page 187**

  Ask the student to give the definition in question 1 verbally before writing it down. For question 2, the examples given are taken from the specification. The statements from question 3 will be used again in the final activity. Discuss why the modification is done at an early stage in development.

- **Cloning; 15 minutes; page 188**

  Ask the student to complete the boxes one at a time and support them in labelling and adding to the diagrams. There are links here to hormones (rooting powder, fertility hormone given to pedigree cow to produce many eggs). For question 3, it might be useful to remind the student of Dolly the sheep.

- **The process of genetic engineering; 10 minutes**

  Use cut outs of the statement boxes from question 3 of the genetic engineering activity. Stick these in order on separate paper and help the student to draw diagrams to represent each step.

Plenary activity

- **Ten words; 5 minutes**

  Ask the student to describe genetic engineering in ten words or fewer. Repeat for cloning. In each case give one minute of thinking time to come up with their answer. They should aim for as much detail as possible using the words available.

Homework activity

- **Using cloning; 15 minutes; page 189**

  The homework is two short answer questions about the use of cloning in plants and considers the adult cell cloning process. Both questions build on the main activities.

Support ideas

- **Genetic engineering** Show the student common foods and discuss links with genetic modification, for example, tomatoes (flavour and ripening), cotton (insect resistant Bt cotton), cheese (animal rennet substitute) and soya (herbicide resistance).
- **Cloning** Using a house plant, demonstrate the technique of taking cuttings. Use the same plant to explain how the procedure would be different using tissue culture.

Extension ideas

- **Genetic engineering** Discuss the example of insect resistant Bt cotton in more detail. Focus on the mechanism by which the bacterial gene was inserted, and risks and benefits involved, including the development of insect resistance.
- **Cloning** Discuss the idea of nuclear transfer and recent developments to allow so called 'three parent baby' procedures in the UK to treat human mitochondrial disorders. Use this to emphasise that DNA in the embryo is from the transferred nucleus.

Progress and observations

BIOLOGY HIGHER

Starter activity: GM and cloning crossword

Time 5 mins

Learning objectives
- To understand some key terms related to genetic modification and cloning

Equipment
none

1. Complete the crossword. The answers are all related to genetic modification and cloning.

Across
5. a simple way to make clones of plants (8)
7. a human hormone that can now be made by genetically modified bacteria (7)

Down
1. genetically identical individuals (6)
2. a small circular piece of DNA found in bacterial cells (7)
3. an organism or agent that is used to carry genes into the genome of an organism (6)
4. these are cut out and transferred to another organism in genetic engineering (5)
6. used to isolate genes in genetic engineering (7)

186

BIOLOGY HIGHER

Main activity: Genetic engineering

Time: 15 mins

Learning objectives
- To describe the process of genetic engineering
- To explain the potential benefits and risks of genetic engineering

Equipment
none

1. Write a definition for 'genetic engineering'.

_____

2. Complete the table to summarise the benefits and potential risks of some types of genetic engineering.

	Modification	Benefits and risks
Plant crops	resistance to disease or insect attack	benefits:
		risks:
	higher yield or to produce bigger better fruits	benefits:
		risks:
	GM crop is resistant to herbicides	benefits:
		risks:
Bacterial cells	produce human insulin	benefits:
		risks:
Medical research using human cells	replacing faulty genes	benefits:
		risks:

3. The statements in the boxes below describe the main steps in the process of genetic engineering. Sort the statements into the correct order.

A. The vector is used to insert the gene into the required cells.	B. Enzymes are used to isolate the required gene.
C. Genes are transferred to the cells to be modified at an early stage in their development so the organism develops with the desired characteristics.	D. This gene is inserted into a vector (such as a bacterial plasmid or a virus).

Correct order: _____

Main activity: Cloning   Time 15 mins

Learning objectives
- To describe some methods of cloning in plants and animals

Equipment
- spare paper
- pencil

1. The boxes below contain diagrams of some methods of cloning plants and animals. Add to the diagrams and add arrow and labels to explain these processes.

Cuttings	Tissue culture

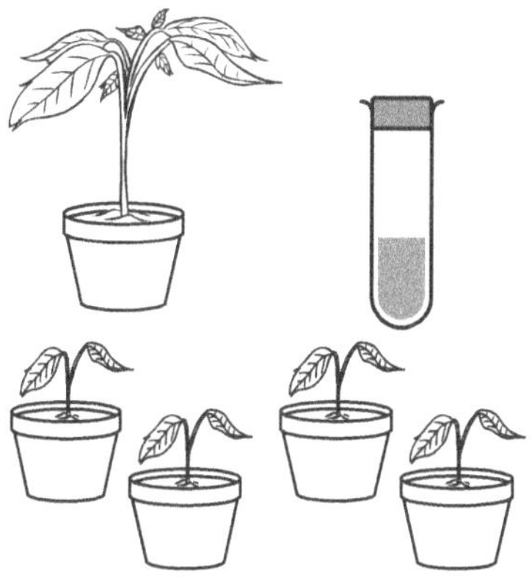

Embryo transplants

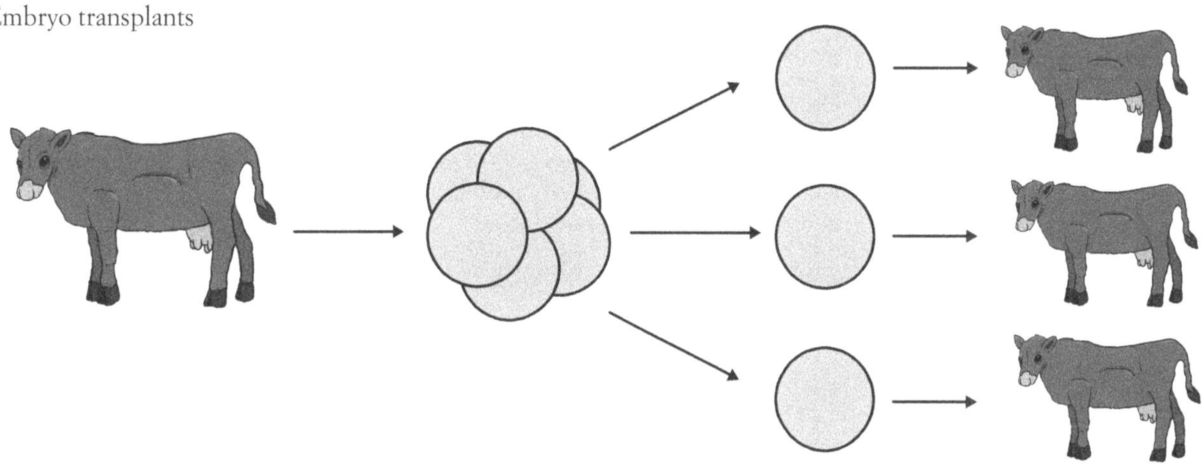

2. On a separate piece of paper draw your own simple diagrams to illustrate these stages of adult cell cloning.

   A. The nucleus is removed from an unfertilised egg cell.

   B. The nucleus from an adult body cell, such as a skin cell, is inserted into the egg cell.

   C. An electric shock stimulates the egg cell to divide to form an embryo.

   D. When the embryo has developed into a ball of cells, it is inserted into the womb of an adult female to continue its development. This embryo contains the same genetic information as the adult skin cell.

BIOLOGY HIGHER

Homework activity: Using cloning

Time 15 mins

Learning objectives
- To understand some uses of cloning

Equipment
none

1. **Two methods of cloning plants include tissue culture and taking cuttings.**

    a) Why is it useful to produce new plants by cloning

    _____

    _____

    _____

    b) Which of the methods, tissue culture or taking cuttings, would be used in the follow situations? Explain your answers.

    i) by gardeners to make new plants from an attractive favourite plant

    _____

    ii) by a botanical garden to preserve a rare plant species from their collection

    _____

    iii) by a commercial nursery to make lots of identical plants for sale

    _____

2. **Scientists wanted to clone a goat that was resistant to a disease. They took an unfertilised egg from a goat called Lisa and removed the nucleus. They then took a cell from the skin of a goat called Maggie. The nucleus from Maggie's skin cell was transferred into Lisa's egg cell. The egg cell was then given an electric shock to make it develop into an embryo. The embryo, now a small ball of cells, was placed into the womb of a goat called Julie. At the end of the pregnancy Julie gave birth to a goat which the scientists called Rose.**

    a) Rose is a clone of one of the adult goats. Which adult goat has cells with the same genetic information as Rose?

    _____

    b) What effect does the electric shock have on the egg cell?

    _____

29 Answers

Starter activity: GM and cloning crossword

1. Across: 5. cuttings 7. insulin
   Down: 1. clones 2. plasmid 3. vector 4. genes 6. enzymes

Main activity: Genetic engineering

1. A process which involves cutting out a gene from one organism and transferring it into the cells of another organism, to give it a desired characteristic
2. Concerns that the effect of eating GM crops on human health have not been fully explored apply to all crop examples.

	Benefits	Risks
Resistant to disease or insect attack	Reduce crop loss; economic benefit; less use of pesticides	May harm wild insects; resistance develops
Higher yield or to produce bigger, better fruits	Economic benefit; more nutritious food	Effects of eating GM crops on human health may not be fully understood
GM crop is resistant to herbicides	Spraying will only kill weeds not crop; reduces competition; economic benefit	Encourages herbicide use; may harm wild plants and reduce biodiversity; gene may transfer to wild plants
Bacteria cells produce human insulin	Treats diabetes; readily available supply; less likely to cause an adverse reaction; not extracted from animals so reduces some ethical concerns	Some people may refuse the drug due to ethical concerns about transferring human genes into other organisms
Medical research using human cells to replace faulty genes	Overcome inherited disorders; reduce human suffering	Germline modifications are passed on to the next generation; possible effects on other genes unknown; possible misuse – 'designer babies'

3. B, D, A, C

Main activity: Cloning

1. Cuttings: include simple method of removing part of stem and leaf, cover in plastic until roots develop, perhaps refer to plant hormone used for rooting
   Tissue culture: show that small group of cells is taken from plant and cultured on sterile agar medium to produce genetically identical new plants
   Embryo transplants: embryos produced using pedigree cow parents, splitting apart cells from a developing (pedigree) embryo before they become specialised, then transplanting the identical embryos into (non-pedigree) host mothers
2. Student's own answers. Ensure the diagrams show the stages of cell cloning in the correct order.

Homework activity: Using cloning

1. a) Many new plants can be produced quickly from one parent plant; they are genetically identical to the parent so they will have the same desirable characteristics; offspring characteristics are not predictable if produced sexually
   b) One mark each for correct answer and explanation
      i) Cutting: simple method; quick; does not need scientific equipment
      ii) Tissue culture: many plants can be made from only one specimen; only small amounts of tissue removed so less damage to parent plant
      iii) Tissue culture: can produce larger numbers of identical plants from single parent than with cuttings
2. a) Maggie
   b) To stimulate the egg cell to divide

30 Inheritance, variation and evolution: Evolution and speciation

Learning objectives

- To describe the work of Darwin and Wallace in the development of the theory of evolution
- To describe the process of speciation by evolution and natural selection

Specification links

- 4.6.2.2
- 4.6.3.1
- 4.6.3.2

Starter activity

- **The who's who of evolution; 10 minutes; page 192**

  Allow the student to complete the activity independently. If they are unsure, provide them with the three names of scientists to match with the information. Afterwards, discuss the work of the three scientists, and mention that Darwin had already written down his ideas years before publishing them but was (quite rightly) worried about the consequences. He was prompted to publish when Wallace came up with the same ideas.

Main activities

- **Theories of evolution; 15 minutes; page 193**

  The student should complete question 1 independently. Ensure that vague answers are corrected and that the three key points – variation, increased survival and passing on of genes – are clear. Question 2 can be done verbally.

  Use examples to support the student's understanding. Question 3 could be used to record key points of a discussion.

- **Speciation; 15 minutes; page 194**

  Ask the student to complete question 1 independently, then check their answers. Question 2 is a series of quick questions with short answers that can be asked verbally.

- **Reproductive isolation; 5 minutes**

  Work together to produce a spider diagram of mechanisms by which two populations may become reproductively isolated.

Plenary activity

- **Two stars and a wish; 5 minutes**

  The student should write down two things that they have learned in the lesson and one question or target for improvement.

Homework activity

- **Island birds; 20 minutes; page 195**

  The homework is a longer exam-style question about speciation.

Support ideas

- **Theories of evolution** Support the student if they cannot give a clear definition of a species by using the example of horses, donkeys and mules.
- **Speciation** Show a short video to explain speciation. An internet search for 'speciation videoclip' will give useful examples.

Extension ideas

- **Theories of evolution** Ask the student to suggest factors that might speed up evolution (such as strong selection pressures, a short life cycle, or a change in environment).
- **Speciation** Show the student pictures of other examples of evolutionary change or speciation such as peppered moths or cichlid fish. Ask the student to explain how the changes in populations may have occurred.

Progress and observations

BIOLOGY HIGHER

Starter activity: The who's who of evolution

Time 10 mins

Learning objectives

- To describe the work of key scientists in the development of the theory of evolution

Equipment

none

1. This page shows information about the works of three famous scientists who were involved in the development of theories about changes in species over time.

   a) Decide who the scientists are and write their name below the relevant picture.

   b) For each scientist, circle the two pieces of information that are most important in the story of how the theory of evolution developed.

- I independently proposed the theory of evolution by natural selection.
- I worked worldwide gathering evidence about evolution.
- I published joint writings about natural selection with another scientist in 1858.
- I am famous for my work on warning colouration in animals and my theory of speciation.

A

- Many people agreed with my ideas at the time.
- I published my ideas about how animals change over time before the other two scientists.
- My main idea was that the environment can change an organism and that such changes can be inherited.
- My ideas were later shown to be mostly incorrect.

B

- My knowledge of geology and fossils helped me with my ideas on natural selection.
- I independently proposed the theory of evolution by natural selection.
- I carried out lots of experiments and discussed my ideas with other scientists for many years before publishing.
- I published joint writings about natural selection with another scientist in 1858. This prompted me to finally publish my own book.
- I made observations on the variation in species during an around the world expedition.
- I published a book of my ideas called On the Origin of Species in 1859.
- I am most famous for proposing the theory of evolution by natural selection.

C

192

BIOLOGY HIGHER

Main activity: Theories of evolution

Time 15 mins

Learning objectives

- To describe the development of the theory of evolution
- To describe the Darwinian theory of evolution by natural selection
- To explain why this theory was only gradually accepted

Equipment

none

1. Read the first box and complete the second box.

Theories about evolution in the early 19th century, including those of **Jean-Baptiste Lamarck**, were based mainly on the idea that changes which occur in an organism during its lifetime can be inherited.
Giraffes have long necks, an adaptation that helps them to survive by feeding on the leaves of trees.

Lamarck's explanation:
Early giraffes had short necks.
When food was short, they had to stretch to reach leaves on trees.
This stretching made their neck a little bit longer.
This acquired characteristic was passed on to offspring.
After many generations, all giraffes had long necks.

Darwin and Wallace's explanation:
Write in the missing steps.
a) Early giraffes had short necks.

b)

c)

d)

e) After many generations, all giraffes had long necks.

2. Charles Darwin proposed the theory of evolution by natural selection. Explain to your tutor what is meant by the terms:

a) natural selection   b) evolution   c) speciation   d) species

3. When Darwin published his theory of evolution by natural selection his ideas were controversial and received some angry reactions. Add to the table to explain why his ideas were only gradually accepted.

Reason	Explanation
Religion	
Evidence	
Understanding	

Main activity: Speciation Time 15 mins

Learning objectives
- To describe the process of evolution by natural selection
- To describe the steps which give rise to new species

Equipment
none

1. Fill in the gaps in this description of how a new species forms. Use the word bank to help you.

All species of living things have evolved over many millions of years from _____ early life forms. These early life forms first appeared on Earth more than three _____ years ago. New species developed from these early life forms by the process of evolution. Evolution relies on the _____ variation within a population, which causes variation in individual phenotypes. Some phenotypes may be better suited to their _____ than others, especially if there is a change in conditions. These better adapted individuals will be more likely to survive and _____ . This is known as _____ . Over many generations, these inherited characteristics become more common in the population. If two populations of one species become so different in phenotype that they can no longer interbreed to produce _____ offspring, they have formed two new species. Speciation is more likely when populations become _____ from each other.

| Word bank: | environment | billion | simple | fertile | isolated | natural selection | genetic | reproduce |

2. Answer the short questions about speciation.

a) What type of organisms were the early life forms mentioned above? _____

b) In Darwin's day the earth was thought to be about 6000 years old. It is now thought to be about 4.5 billion years old. Where does the modern evidence for the age of the earth come from?
_____

c) How does genetic variation within a population arise? _____
_____

d) What sort of change in conditions might cause a rare phenotype to be selected?
_____

e) How might two populations become isolated from each other? _____
_____
_____

194

BIOLOGY HIGHER

Homework activity: Island birds

Time 20 mins

Learning objectives
- To describe the process of speciation by evolution and natural selection

Equipment
none

1. Darwin studied small birds called finches in the Galapagos islands. There are several similar but different species of finch on different Galapagos islands. The information below is a simple version of what he found.

Bird A is the type of bird that is thought to have originally arrived on the islands from the mainland many thousands of years ago.
Species B and C are found on two different islands.
Suggest an explanation for the development of these different species, B and C.

_____
_____
_____
_____
_____
_____
_____
_____
_____
_____
_____
_____
_____
_____
_____
_____
_____

BIOLOGY HIGHER

30 Answers

Starter activity: The who's who of evolution

1. a) A. Alfred Wallace, B. Jean-Baptiste Lamarck, C. Charles Darwin
   b) Discuss the student's ideas.

Main activity: Theories of evolution

1. b) Individual organisms within a particular species show a wide range of variation for a characteristic.
   c) Individuals with characteristics most suited to the environment are more likely to survive to breed successfully.
   d) The characteristics that have enabled these individuals to survive are then passed on to the next generation.
2. See glossary definitions on page 245
3.

Reason	Explanation
Religion	The theory challenged the idea that God made all the animals and plants that live on Earth.
Evidence	There was insufficient evidence at the time the theory was published to convince many scientists.
Understanding	The mechanism of inheritance and variation was not known until 50 years after the theory was published.

Main activity: Speciation

1. Simple, billion, genetic, environment, reproduce, natural selection, fertile, isolated
2. a) Single-celled organisms/bacteria
   b) Geology, radioactive analysis of rocks
   c) mutations
   d) Change in climate, predators, food or habitat
   e) Geographic barrier such as sea, mountain range, river

Homework activity: Island birds

1. Student's own answer. Any five from: the populations are separated by geographical barrier/the sea; there is variation in the populations; idea of genetic differences/alleles/mutations; there is different food available/different environmental conditions on each island; natural selection for the beak shape best suited to conditions on each island; birds with best-adapted beak more likely to survive to breed; different allele(s) passed on to offspring in different environments/islands; differences become so great that they can no longer interbreed

BIOLOGY HIGHER

31 Inheritance, variation and evolution: Evidence for evolution – extinction

Learning objectives

- To describe the evidence for evolution provided by fossils and antibiotic-resistant bacteria
- To describe factors which may contribute to the extinction of a species

Specification links

- 4.3.6.4
- 4.3.6.5
- 4.3.6.6
- 4.3.6.7

Starter activity

- **Evidence for evolution; 5 minutes; page 198**

  The student should complete the activity. Question 1 explores the student's understanding of fossils and question 2 covers an overview of the evidence for evolution.

Main activities

- **Fossils; 15 minutes; page 199**

  The student may need some additional explanation of the processes and issues outlined in questions 1 and 2. Complete question 1 in collaboration with the student. Questions 2–5 can be answered verbally.

- **Antibiotic-resistant bacteria; 15 minutes; page 200**

  Make sure that the student understands the initial diagrams, then ask them to complete questions 1 and 2 on the sheet. Questions 3–5 can be done verbally. A common misunderstanding is that antibiotics cause mutation rather than select for mutated strains. Make sure the student is aware that this is not the case. Question 6 considers steps to prevent resistance developing.

- **Causes of extinction; 10 minutes**

  The student should be able to describe factors which may contribute to the extinction of a species. Show a series of pictures as prompts for different factors, ask the student to suggest what they represent and write a bullet point list (for example, asteroid impact, disease, super volcanoes, habitat loss, loss of food source, competition, over-predation, invading species and climate change). Factors may cause global extinctions. Ask the student to suggest which factor may cause the biggest impacts. Emphasise that although habitat loss events may individually be small and localised, they are a major cause of extinctions at present.

Plenary activity

- **Convince me; 5 minutes**

  Tell the student that you will play the part of a scientist in Darwin's time who did not agree with the idea of evolution by natural selection. The student has two minutes to convince you of evolution by explaining the available evidence.

Homework activity

- **All about evolution; 45 minutes; page 201**

  The student is asked to produce a revision resource that includes ideas about evolution covered over the last two lessons.

Support ideas

- **Fossils** Show a short video or animation of how fossils can form. Imprints can be modelled using Plasticine®. Show pictures of fossilised humans preserved in peat with conditions too acidic for decay, or frozen mammoths from glaciers.
- **Antibiotic-resistant bacteria** Emphasise that antibiotics work alongside the body's own defences.

Extension ideas

- **Fossils** Show a variety of examples of evolutionary trees. Ask the student questions about the relationships between organisms, such as their most recent common ancestor.
- **Antibiotic-resistant bacteria** Look at the first picture of bacteria on the sheet. Ask the student why the numbers of any resistant bacteria would remain low or die out if antibiotics were not used (they would be outcompeted by non-resistant strains).

Progress and observations

BIOLOGY HIGHER

Starter activity: Evidence for evolution Time 5 mins

Learning objectives
- To describe the evidence for evolution

Equipment
none

1. Darwin knew that fossils provided evidence for evolution. He compared the fossils in newer and older rocks.

   a) What did the fossil record show about the types of species in new rocks compared with older rocks?

   _____

   b) Which layers of rocks are usually (but not always) the oldest?

   _____

2. The theory of evolution by natural selection is now widely accepted. Darwin would be pleased to know that there has been a lot more evidence for evolution since he first proposed his theory.

   Think how you would explain the new evidence to Darwin. Write a few words that you would mention if you could speak to him today, under each heading.

   Fossils

   _____
   _____
   _____

   Medical treatment and bacteria

   _____
   _____
   _____

   Genes

   _____
   _____
   _____

BIOLOGY HIGHER

Main activity: Fossils   Time 15 mins

Learning objectives

- To describe the evidence for evolution provided by fossils
- To describe factors which may contribute to the extinction of a species

Equipment
none

1. Fossils are the remains of organisms from millions of years ago, which are found in rocks. There are three main ways in which fossils may be formed. These are represented by the diagrams below. Suggest to your tutor what type of fossilisation each of these diagrams represents. Write a few words next to each picture to record your ideas.

2. Scientists cannot be certain about how life began on Earth. This is because of problems with the fossil record. Suggest to your tutor what some of these problems might be.

3. Many of the plants and animals that are found as fossils no longer exist. How do we describe such organisms?

4. Explain why fossils of vertebrates often just look like skeletons of the organism.

5. Sometimes, fossils of marine organisms are found in rock on the tops of mountains. How does this happen?

BIOLOGY HIGHER

Main activity: Antibiotic-resistant bacteria

Time: 15 mins

Learning objectives
- To explain how antibiotic-resistant strains of bacteria evolve
- To explain some methods of reducing the development of resistant strains

Equipment
- pencil

Mutations occur continuously in the DNA of bacteria. This produces new strains, some of which might be resistant to antibiotics. Harmful bacteria invade a person's body and reproduce quickly. The person's doctor gives them antibiotics to fight the pathogen. Antibiotics may not kill all the bacteria. Any bacteria that are resistant to the antibiotic will not be killed, so more of these will survive than the non-resistant strain.

1. The black dots in the diagram represent the antibiotic-resistant bacteria. The white dots represent the non-antibiotic-resistant bacteria. In the box, draw what happens when each of the surviving bacteria reproduce, after two divisions.

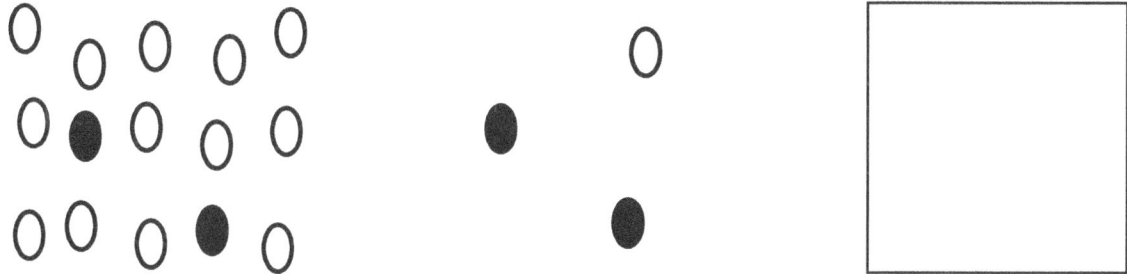

2. What will happen to the population of the resistant bacteria in the person?

_____

3. The resistant strain of the bacteria will spread rapidly in the nearby human population. Why does this happen?

_____

4. Why do antibiotic resistant strains of bacteria evolve so rapidly?

_____

5. MRSA is resistant to antibiotics. Tell your tutor two other facts about MRSA.

6. The development of new antibiotics is costly and slow. Medicine is losing the battle against the emergence of new strains of antibiotic-resistant bacteria. Complete the boxes to explain what each group can do to reduce the rate of development of resistant strains.

Doctors	
Patients	
Farmers	

200

Homework activity: All about evolution

Time 45 mins

Learning objectives
- To be able to explain the theory and evidence for evolution

Equipment
- spare paper
- pencil
- coloured pencils or pens

1. Produce a revision resource to summarise all you have learned about evolution.

   Include information of the history of development of the theory of evolution, how the process works and the evidence for evolution. Produce either revision cards, a spider diagram or a revision poster. Use colour and diagrams where possible.

31 Answers

Starter activity: Evidence for evolution

1. a) It showed how much species had changed over time; there are new species in newer rocks and species from older rocks had become extinct
   b) Lower/deeper layers are usually older
2. Fossils: more fossils have been found, giving better evidence of changes over time; there are also better methods of dating rocks so we know how old the fossils are
   Bacteria: we now have knowledge of how resistance to antibiotics evolves in bacteria, an example of evolution
   Genes: it has been shown that characteristics can be passed on to offspring; genes provide the mechanism

Main activity: Fossils

1. A: Parts of organisms do not decay because one or more of the conditions needed for decay are absent. B: Parts of the organism are replaced by minerals as they decay. C: Traces of organisms are preserved, such as footprints, burrows and rootlet traces.
2. Many early forms of life were soft-bodied, so they have left few traces behind. What traces there were have been mainly destroyed by geological activity.
3. extinct
4. The soft parts of an organism are not preserved well, but the hard bones last long enough to fossilise.
5. The Earth's tectonic plates are constantly moving; mountain ranges may be pushed up when they collide; also sea level may have changed

Main activity: Antibiotic-resistant bacteria

1. Diagram with 8 resistant and 4 non-resistant bacteria.
2. The population of the resistant strain rises.
3. The resistant strain will spread because people are not immune to it and there is no effective treatment.
4. Bacteria can evolve rapidly because they reproduce at a fast rate.
5. Any relevant detail, for example: symptoms, problems of bacteria spreading in hospital, vulnerable individuals, detail of antibiotic resistance, and so on
6.

Doctors	Should not prescribe antibiotics inappropriately, for example, not for non-serious or viral infections
Patients	Should complete any course of antibiotics so all bacteria are killed and none survive to mutate and form resistant strains, so resistant strains are not selected
Farmers	Agricultural use of antibiotics should be restricted

Homework activity: All about evolution

1. Student's own answer. Check detail against specification for missing sections and discuss how the resource could be used for revision.

BIOLOGY HIGHER

32 Ecology: Classification and communities

Learning objectives	Specification links
• To understand the development of classification systems	• 4.6.4
• To explain the importance of interdependence and competition in a community	• 4.7.1.1

Starter activity

- **Ecology words; 5 minutes; page 204**

  Provide the student with a set of the words and definitions which have been cut into boxes. Ask them to match the words to the definitions.

Main activities

- **Classification; 15 minutes; page 205**

  This activity covers the Linnaean classification system and more recent developments, along with evolutionary tree diagrams. Ask the student to complete question 1 independently, then check their understanding before continuing. Questions 2 and 4 can be completed as a discussion. Questions 5 and 6 can be answered on the sheet.

- **Communities and competition; 15 minutes; page 206**

  This activity examines ideas about interdependence and competition in communities which should not cause difficulties for most students. Question 2 is to be completed in the table; other questions can be done verbally if time is short. The terms interspecific and intraspecific (question 4) do not feature in the specification but may be useful terms. Adaptations are the focus of lesson 33.

- **Woese's story; 10 minutes**

  Provide the student with the background story of Woese's observations and the history of their acceptance in more detail (using RNA, published in 1977 in a top journal, not accepted by scientists until 20 years later). Ask the student to suggest why acceptance was slow (contradicted current understanding, not accepted until more evidence backed it up, and so on).

Plenary activity

- **What's the word?; 5 minutes**

  Give the student the cards (without definitions) from the *Ecology words* activity. The student should describe each thing in turn without saying the keyword. The tutor must guess the word when the explanation is clear and they 'bank' the word card if they are correct. The student should aim to bank as many of the word cards as possible in the time available.

Homework activity

- **Ecosystems and evolutionary trees; 20 minutes; page 207**

  A series of exam-style questions about organisation within ecosystems and interpretation of evolutionary tree diagrams.

Support ideas

- **Classification** The tiger is in the animal kingdom. What are the other four kingdoms? Ask the student to outline the features of each.
- **Communities and competition** To reinforce the idea of interdependence, repeat the exercise in question 2 for a pond or grassland ecosystem, using a picture or diagram as a prompt. Ask the student to suggest links between species for each category in the table.

Extension ideas

- **Classification** Ask the student to suggest a mnemonic to help them to remember the hierarchical grouping of the Linnaean system.
- **Communities and competition** Use the example of Gause's classic 1936 experiment using two species of algae growing in culture. Ask the student to explain what they compete for and why one might outcompete the other.

Progress and observations

  **BIOLOGY HIGHER**

Starter activity: Ecology words — Time 5 mins

Learning objectives
- To understand some of the keywords in ecology
- To describe different levels of organisation in an ecosystem

Equipment
- cut out set of the boxes below to make sort cards

1. Your tutor will provide you with a set of sort cards. Match the words with the definitions.

adaptation	physical or behavioural features that allow a species to be well suited to and survive in its environment
abiotic factor	a non-living condition that can influence the organisms in an ecosystem
biotic factor	a condition caused by living things that can influence the organisms in an ecosystem
biomass	the mass (usually when dried) of a population or community of living things in an area
community	all the animals and plants living in a particular area
competition	when organisms have a damaging effect on each other as a result of needing the same limited resource
consumer	living thing that gets its energy from other living things, usually by eating them
ecology	the study of living things in their environment
ecosystem	all the living things combined with the physical/chemical conditions in an area
environment	the physical and chemical conditions surrounding an organism
habitat	the specific place where an organism lives
population	the number of individuals of one species in an area
predator	animal that captures and eats other animals
prey	animal that is captured and eaten by other animals
producer	organism that makes complex organic nutrients from simple molecules using sunlight (photosynthesis) or chemical energy (chemosynthesis)

BIOLOGY HIGHER

Main activity: Classification   Time 15 mins

Learning objectives
- To understand the Linnaean classification system
- To describe the impact of developments in biology on classification systems

Equipment
none

1. The formal system of classifying organisms into groups was developed in the 1700s by Carl Linnaeus, a Swedish scientist. This system uses differences in structure and characteristics to place organisms into groups. The classification system is hierarchical, with organisms gradually split into smaller and smaller groups. Fill in the gaps in the table to show the missing group names in this example for the tiger.

Group	Example	Explanation
kingdom	*animalia*	
	chordata	vertebrates
class	*mammalia*	
	carnivora	carnivores
	felidae	cats
genus	*Panthera*	big cats
species	*tigris*	tiger

2. Suggest to your tutor what sort of characteristics might have been used to separate the tiger from the lion which is a closely related species.

3. Organisms are named by the binomial system using two names taken from this classification system. Which two are used? Spell to your tutor the correct binomial name for the tiger.

4. Over time, improvements in technology have allowed better evidence to be used to separate organisms into groups. Suggest to your tutor how improvements in microscopes may have improved classification.

5. As technology improved and evidence from chemical analysis became available, new models of classification were proposed. Carl Woese analysed the genetic material of living things and formed the three domain system. Draw lines to join the domain name to the correct description.

| archaea | | includes protists, fungi, plants and animals |

| bacteria | | primitive bacteria, usually living in extreme environments |

| eukaryota | | true bacteria |

6. Evolutionary trees' are diagrams of different sorts used by scientists to show how organisms are related, based on the evidence available. The diagram shows a simple evolutionary tree for the three domains.

   a) Circle the domain that shares the most recent common ancestor with the eukaryota.

   b) What point on the tree shows the divergence of all groups from a single common ancestor?

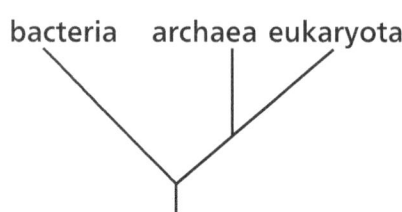

BIOLOGY HIGHER

Main activity: Communities and competition   Time 15 mins

Learning objectives

- To explain the importance of interdependence and competition in a community
- To explain how organisms are adapted to the conditions in which they live

Equipment

none

The diagram represents a woodland ecosystem which includes living organisms such as oak trees, bluebells, foxes, rabbits, fungi, squirrels and bees.

1. Apart from organisms, what else forms a part of any ecosystem?

_____

2. Interdependence between species is important within a community. Find examples from the diagram and complete the table.

Type of interdependence	Example from woodland
seed dispersal	
food	
pollination	
shelter	
recycling of mineral ions	

3. Explain to your tutor what would happen to the ecosystem if a) the oak trees or b) foxes were removed.

4. When organisms need the same resources, they interact by competing with each other. For the woodland example, explain what resources the following species would compete for:

   a) bluebells and oak trees

   b) two squirrels of the same species

   c) What types of competition are represented in a) and b)?

5. Bluebells are one of the earliest plants to grow and flower in the spring. They appear long before the deciduous oak tree regrows its leaves. Explain why this is an adaptation to living in the wood.

6. Suggest one adaptation that the squirrel has for living in the woodland ecosystem.

7. Oak woods are usually stable ecosystems. Explain to your tutor what is meant by a 'stable ecosystem'.

206

BIOLOGY HIGHER

Homework activity: Ecosystems and evolutionary trees   Time 20 mins

Learning objectives
- To describe different levels of organisation in an ecosystem

Equipment
none

1. Words describing four levels of organisation in ecology are:

   A. population   B. ecosystem   C. organism   D. community

   Write a letter in the table to match these words to the correct definition.

	an individual living thing
	the number of individuals of one type of organism living in an area
	all the animals and plants living and interacting in a particular area
	the living organisms and the non-living characteristics of an area and the interactions between them

2. Living things within a community may compete for resources. List in the table the types of factors that plants or animals might compete for.

Plants	Animals
1.	1.
2.	2.
3.	3.
4.	

3. This evolutionary tree diagram shows the relationships between four species of snail, based on evidence that a group of scientists collected.

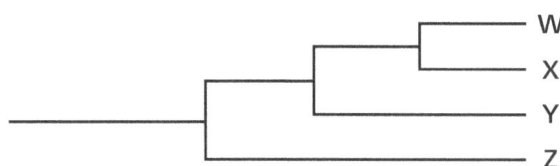

   a) Which snail is most closely related to snail X?
   _____

   b) What do scientists mean when they describe two organisms as being 'closely related'?
   _____

   c) What is represented by the branching of the diagram?
   _____

   d) Snail Z is extinct while the other species still exist today. Describe the evidence that scientists could use to describe the relationships between both living organisms and extinct organisms.
   _____

32 Answers

Starter activity: Ecology words

1. Answers on the intact starter activity sheet are correct.

Main activity: Classification

1.

Group	Example	Explanation
kingdom	*animalia*	animals
phylum	*chordata*	vertebrates
class	*mammalia*	mammals
order	*carnivora*	carnivores
family	*felidae*	cats
genus	*Panthera*	big cats
species	*tigris*	tiger

2. Body shape, head shape, fur patterns and colour, relative length of parts of body
3. *Panthera tigris*, must have capital letter for genus, underline or italics to show words are in Latin
4. Differences or similarities in internal structures could now be used in classification and small organisms could be better classified.
5. Archaea: primitive bacteria, usually living in extreme environments; bacteria: true bacteria; eukaryota: includes protists, fungi, plants and animals
6. a) archaea
   b) Node nearest the root

Main activity: Communities and competition

1. Non-living (abiotic) parts
2. Multiple examples, such as seeds: squirrel–oak; food: rabbit–fox; pollination: bee–bluebell; shelter: squirrel–oak; recycling: fungi–oak tree
3. Various possible answers, for example:
   a) Loss of habitat/food for squirrel-numbers would decrease, bluebells might increase then be replaced by species better adapted for new conditions
   b) Rabbit numbers would increase, oak tree numbers might decrease due to saplings being eaten
4. a) Bluebell/tree: light, space, water, mineral ions
   b) Squirrels: food, mates, territory
   c) Interspecific and intraspecific
5. The oak and bluebell compete for light. By growing from bulbs early in the season, bluebells can photosynthesise and gain energy for growth/reproduction before the oak tree grows its leaves and reduces light levels on the forest floor.
6. Long tail for balance when living in trees, sharp teeth to break open acorns, any other valid adaptation
7. One where all the species and environmental factors are in balance so that population sizes remain fairly constant

Homework activity: Ecosystems and evolutionary trees

1. C, A, D, B
2. Plants: light, space, water, mineral ions; animals: food, mates, territory
3. a) W
   b) They shared a recent common ancestor; they have many similar characteristics
   c) Branching occurs when one species splits into two species; new species formed by natural selection/evolution
   d) Any three from: they use data from fossils for extinct organisms; they look at differences in structure for both extinct and living organisms; for living organisms, chemical analysis can also be used such as, differences in DNA/RNA/proteins

BIOLOGY HIGHER

33 Ecology: Adaptations, abiotic and biotic factors

Learning objectives

- To explain how changes in abiotic or biotic factors affect communities
- To explain how organisms are adapted to the conditions in which they live

Specification links

- 4.7.1.2
- 4.7.1.3
- 4.7.1.4
- MS 2d, 4a

Starter activity

- **Desert plants; 5 minutes; page 210**
  Allow the student to complete the activity independently. When discussing answers ask the student which challenges are biological/non-biological and introduce the terms biotic/abiotic. Ask them to name the term that describes features which help survival (i.e. adaptations).

Main activities

- **Abiotic and biotic factors; 15 minutes; page 211**
  Allow the student to answer question 1 independently then review their answers. Question 2 provides a graph and questions as a focus for discussion. Some explanation of river communities may be needed.
- **Adaptations; 15 minutes; page 212**
  Introduce the first activity, then provide time for the student to fill in the table independently. Some explanation of the rocky shore environment and tides might be needed. The existence of extremophile bacteria at deep sea vents is required knowledge for the specification; the second section covers this topic; questions can be completed verbally.
- **Measuring abiotic factors; 10 minutes**
  For each of the abiotic factors identified in the table in the first main activity, ask the student to suggest how they would monitor or measure the factor (for example light meter, oxygen meter or titration methods, chemical colorimetric tests for minerals, and so on). Ask how often they would take measurements and emphasise the importance of measuring to pick up seasonal and diurnal patterns.

Plenary activity

- **Highlights; 5 minutes**
  Ask the student to highlight the five most important keywords or phrases from the lesson. Provide sufficient thinking time for this. They should then explain to you why these phrases are the most important to the topic.

Homework activity

- **Changes in abiotic and biotic factors; 15 minutes; page 213**
  The homework is short, exam-style questions based around interpretation of a table of data.

Support ideas

- **Abiotic and biotic factors** Show the student pictures of river organisms to help visualisation. An internet search for 'river food-web diagram' will give useful results.
- **Adaptations** Students who have not seen images of the communities at deep sea vents may fail to understand the significance and organisation of this ecosystem. An internet search for 'hydrothermal vent video clip' will be useful.

Extension ideas

- **Abiotic and biotic factors** Bloodworms are red because they contain haemoglobin in their bodies. Ask the student to suggest why this is an adaptation to low oxygen levels.
- **Adaptations** Ask the student to suggest what advantages the ability to survive in the intertidal zone gives the limpet in terms of survival – what was the evolutionary pressure to leave the sea? Discuss the idea that adapting to extreme abiotic environments allow organisms to exploit areas where biotic factors such as predation and competition are lower.

Progress and observations

BIOLOGY HIGHER

Starter activity: Desert plants

Time 5 mins

Learning objectives
- To identify factors that affect survival and features that help organisms to survive

Equipment
none

1. The diagram shows some plants that live in a desert.

a) What challenges for survival do the plants face both from their environment and from other plants? List your ideas.

_____
_____
_____

b) What features do the plants have that help them to survive? List your ideas.

_____
_____
_____

BIOLOGY HIGHER

Main activity: Abiotic and biotic factors Time 15 mins

Learning objectives
- To describe a range of abiotic or biotic factors
- To interpret graphs relating to the effects of abiotic or biotic factors

Equipment
none

1. There are many factors that can affect a community. Some of these factors are living (biotic) and others are non-living (abiotic). Place a tick (✓) in the correct column to show whether the following factors are biotic or abiotic.

Factor	Abiotic	Biotic
light intensity		
oxygen levels (for aquatic animals)		
competition		
soil mineral content		
moisture levels		
temperature		
appearance of a new type of pathogen		
wind strength or direction		
new predators arriving in the area		
carbon dioxide levels (for plants)		
soil pH		
availability of food		

2. Human and animal waste decays in water. Bacteria break down the organic material in the waste by aerobic respiration. If a sewage treatment system is not working properly, any waste entering a river can cause pollution. The graph shows the results of a study where samples of invertebrate populations were taken both upstream and downstream from an outfall where sewage waste entered the river.

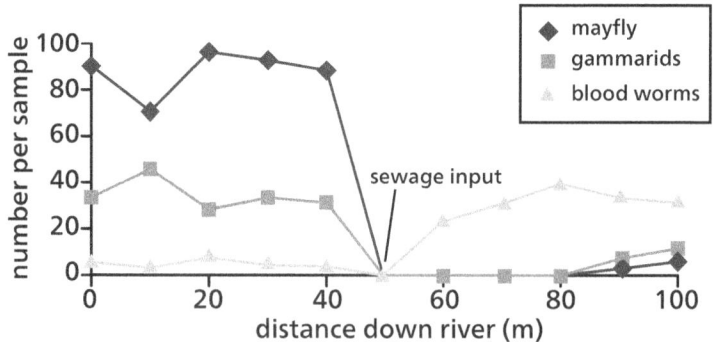

   a) Explain why there are no invertebrates at 50 m along the sampling section of river.

   _____
   _____

   b) Suggest why populations of bloodworms dominate the community downstream of the sewage input.

   _____
   _____

   c) Explain why bloodworm numbers are low where water is unpolluted (above the sewage outfall).

   _____

   d) The sewage adds organic material to the water and makes it cloudy. There can be positive and negative implications for aquatic plants. What associated changes in abiotic factors might affect plant growth?

   _____

BIOLOGY HIGHER

Main activity: Adaptations   Time 15 mins

Learning objectives

- To explain how organisms are adapted to live in their environment
- To describe adaptations as structural, behavioural or functional
- To describe deep sea bacteria as an example of extremophiles

Equipment

- video or pictures of deep sea vent communities

1. Adaptations are the features that organisms have that enable them to survive in their environment. Adaptations may be structural, behavioural or functional. Read the description of the limpet and write down examples of adaptations under each heading in the table.

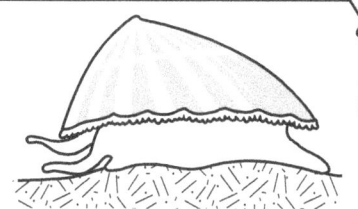

The common limpet lives on rocky shores in the intertidal zone. This means that as the tide comes in and out twice a day, the limpet is first covered in water and then exposed to the air for several hours at a time. The limpet is a marine snail and has gills inside its shell that can only function when wet, so it has evolved features that allow it to survive out of water. It has a waterproof shell that grows to fit the surface of its home location perfectly, making a water-tight seal when the tide is out and using its strong, muscular 'foot' to pull it firmly to the rock. It moves around to feed on algae only when the tide is in. It also reduces water loss by clamping down in more humid areas when the tide is out, such as in cracks, under seaweed, or in limpet 'huddles'. The limpet has developed body processes that allow it to tolerate a wider range of temperatures than most marine organisms. It can also cope with a greater loss of water from its body and a wider range of salinities than most organisms.

Structural	Behavioural	Functional

2. Some organisms live in very extreme environments. Bacteria living in deep sea vents are one example. These bacteria live kilometres under the surface of the sea, at deep sea vents, where very hot water (sometimes over 700 °C) leaves the Earth's crust.

   a) What is the name for organisms that live in extreme environments? _____

   b) Besides high temperatures and dark conditions, what other extreme abiotic factors are a feature of the deep sea vent ecosystem?

   _____

   c) The cellular processes of bacteria living in hot water environments rely on enzymes made of special proteins. These proteins denature at much higher temperatures than those in most organisms. What type of adaptation is this?

   _____

   d) The bacteria are producers in an ecosystem that receives no sunlight, so they can't photosynthesise. How do these bacteria gain energy for themselves (and through food chains for the whole ecosystem)?

   _____

212

BIOLOGY HIGHER

Homework activity: Changes in abiotic and biotic factors

Time 15 mins

Learning objectives

- To interpret information from tables relating to the effect of abiotic or biotic factors on organisms within a community

Equipment

none

1. A survey was carried out on the abundance of wild flowers in a field, before and after construction of a new road passing through the field. Some abiotic factors were also measured. Plant abundance and abiotic factors were measured at 30 different sites in the field. The mean and range were then calculated for each type of measurement.

		Wild flower abundance	pH	soil moisture content	Soil nitrate content
		m^{-2}		%	$mg\ kg^{-1}$
Before construction works	mean	15	6.5	36	10
	range	11–18	5.6 - 7.5	30–38	8–13
After construction works	mean	3	6.3	18	11
	range	1–6	5.5 – 7.5	15–20	8–13

a) Describe any change in the mean abundance of wild flowers after construction of the road.

_____
_____
_____

b) A report stated that measurements of abiotic factors had changed following construction. Write a more detailed description of the results shown in the table.

_____
_____
_____

c) The field was monitored again five years later. It was found that the original species of flower had disappeared and a new species had taken its place. Suggest why this may have happened.

_____
_____
_____

33 Answers

Starter activity: Desert plants

1. a) Lack of water, high temperatures, competition for water from other plants
   b) Shallow/extensive root systems, storage of water in stem, thick waxy cuticle/leaves reduced to spines so less transpiration/water loss

Main activity: Abiotic and biotic factors

1.

Factor	Abiotic	Biotic
Light intensity	✓	
Oxygen levels (for aquatic animals)	✓	
Competition		✓
Soil mineral content	✓	
Moisture levels	✓	
Temperature	✓	
Appearance of a new type of pathogen		✓
Wind strength or direction	✓	
New predators arriving in the area		✓
Carbon dioxide levels (for plants)	✓	
Soil pH	✓	
Availability of food		✓

2. a) Oxygen levels are too low for life (needed for aerobic respiration)
   b) Bloodworms are better adapted to survive in the low oxygen environment than the other species.
   c) Competition from mayflies and gammarids keeps numbers low, as these species are better adapted to clean water.
   d) Cloudy so less light penetrates water, photosynthesis reduced; organic matter adds mineral ions, increases growth/acts like fertiliser

Main activity: Adaptations

1. Structural: waterproof shell, shell shape makes seal with rock, muscular foot; behavioural: choice of humid home site, only move/feed when under water, clamp down when tide is out; functional: can tolerate desiccation and a wide range of temperatures and salinities
2. a) extremophiles
   b) High pressure and toxic chemicals
   c) functional
   d) Chemosynthesis: they gain energy by oxidising chemical substances in the vent water

Homework activity: Changes in abiotic and biotic factors

1. a) Abundance of wild flowers is much lower after construction works; the mean abundance is only one fifth of/20% of the abundance before construction/less than the minimum abundance previously
   b) Any three points from: only soil moisture content has changed significantly/little change in pH and nitrate; soil moisture decreased to 50%/one half of its former value; ranges of the soil moisture values before and after construction do not overlap; ranges of pH and nitrate are very similar before and after construction
   c) The new species was better adapted to the drier soil conditions; the new species outcompeted the original species; eventually the numbers of the original species were not sufficient to breed and they were no longer found in the field

BIOLOGY HIGHER

34 Ecology: Organisation, cycling of materials and decomposition

Learning objectives

- To understand that many different materials cycle through the abiotic and biotic components of an ecosystem
- To explain the process, rate and importance of decomposition
- To understand how to sample populations and interpret predator–prey models

Specification links

- 4.7.2.1
- 4.7.2.2
- 4.7.2.3
- MS 2b, 2d, 2f

Starter activity

- **Food in a chain, water in a cycle; 5 minutes; page 216**

  There is a lot to do in the time but the content should be easy, so encourage the student to move through the tasks quickly.

Main activities

- **Populations, predators and prey; 15 minutes; page 217**

  The first section is focused on required practical 9. Questions should be answered as a discussion, but notes and annotations of the diagram can be made by the student. The second part looks at predator–prey cycles, which should not be challenging for most GCSE students.

- **Cycling of materials and decomposition; 15 minutes; page 218**

  Ask the student to complete the boxes and label arrows on the carbon cycle. Check their answers, and emphasise the importance of microorganisms in returning carbon to the atmosphere. The second part covers the required practical on measuring the rate of decay of fresh milk by measuring pH change. The student will not be able to complete all the detail in the time available but the table should be used to record key points.

- **Using decomposition; 10 minutes**

  Ask the student to draw a Venn diagram on separate paper with the headings 'gardeners' compost' and 'biogas'. They should then record factors in common in the overlapping portion such as decay, factors that speed up decay, and write differences in the remaining space such as aerobic/anaerobic, uses of products and so on.

Plenary activity

- **Journey of a carbon atom; 5 minutes**

  Ask the student to describe the journey of a carbon atom as it travels from being part of a lipid molecule in a dead rabbit to being part of a cellulose molecule in an oak tree. After they have made some notes, ask them to talk through the journey. Prompt them for names of processes such as decomposition, photosynthesis and synthesis.

Homework activity

- **Decay and sampling; 30 minutes; page 219**

  The student should complete the short exam-style questions which include calculations of mean, median and population.

Support ideas

- **Populations, predators and prey** If the student has had little experience of using a quadrat, use a sheet of newspaper and a 'quadrat' made from acetate or wire. Discuss how to sample to estimate the population of a letter on the page.
- **Cycling of materials and decomposition** If support is needed for the carbon cycle, show or sketch a simplified diagram with only plant, animal, soil and atmosphere. Focus only on the key specification points.

Extension ideas

- **Populations, predators and prey** Ask the student to explain how they could find out how many quadrats are sufficient (running mean). Then ask them to explain why predator numbers are always much lower than prey numbers.
- **Cycling of materials and decomposition** Ask the student to sketch a graph to predict the relationship between temperature and rate of decay of milk.

Progress and observations

BIOLOGY HIGHER

Starter activity: Food in a chain, water in a cycle

Time 5 mins

Learning objectives
- To describe the levels of organisation in a food chain
- To explain the importance of the water cycle

Equipment
- pencil

1. Complete the sentences using the word bank to help.

Feeding relationships within a community can be represented by food _____ . All food chains begin with a _____ which synthesises molecules. This is usually a green plant or _____ which makes _____ by photosynthesis. These photosynthetic organisms are the producers of biomass for life on Earth. Producers are eaten by _____ consumers, which in turn may be eaten by _____ consumers and then _____ consumers. Consumers that kill and eat other animals are called _____ , and those eaten are called prey.

| Word bank: | predators | chains | primary | producer | glucose | secondary | alga | tertiary |

2. Rabbits eat grass and are in turn eaten by foxes. Draw a food chain to represent this feeding relationship.

3. Annotate the diagram below with labels and arrows to show the key processes in the water cycle.

216

BIOLOGY HIGHER

Main activity: Populations, predators and prey   Time 15 mins

Learning objectives

- To understand how to use sampling to investigate populations and the effect of a factor on the distribution of a species
- To be able to interpret graphs that model predator–prey cycles

Equipment

none

1. Terry is using a quadrat to investigate the population and distribution of a plant in a field. He observes that the plant seems more abundant in areas that are not shaded by a large tree in the centre of the field. He decides to compare the abundance of the plant in shaded and open areas by placing some quadrats under the tree and some away from the tree. Discuss Terry's work with your tutor using the questions below.

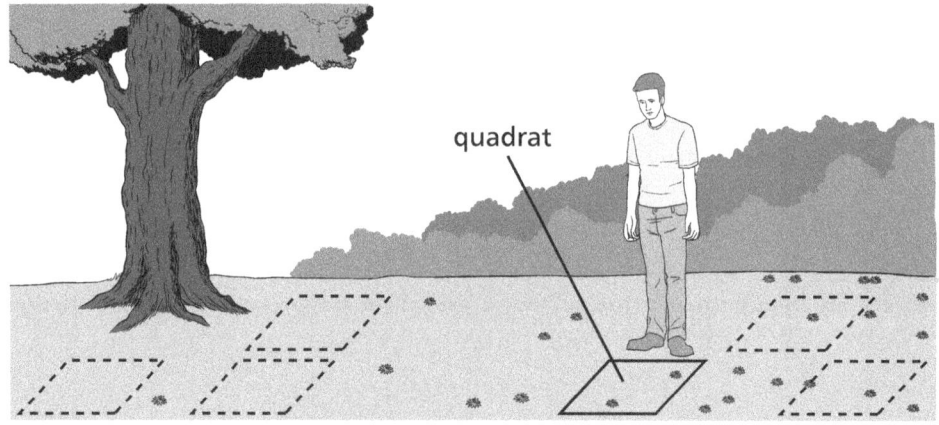

   a) What is meant by the terms abundance, distribution and population?

   b) Using a quadrat is one way of sampling the area. Why does Terry take a sample rather than counting all the plants?

   c) Explain how Terry would use the quadrat to compare the two areas. What would he do about plants that lie across the edge of the quadrat?

   d) Terry is counting the plants in each quadrat. What method would he use to study a plant like moss, where individual plants cannot be distinguished?

   e) Placing quadrats in two locations will allow Terry to compare abundance in the two areas. What would be a better way to investigate the changes in distribution as you move from an area of more shade to an area of less shade? Sketch the arrangement on the diagram.

   f) What abiotic factor should Terry measure and how would he do this?

2. In a stable community the numbers of predators and prey rise and fall in cycles. Add letters to the diagram to show an example for each of the explanations A to D below.

   A. Prey populations survive well and rapidly increase because predator numbers are low.

   B. This is followed by an increase in the predator population because they have lots of food and reproduce quickly.

   C. As predators eat the prey, the prey population decreases.

   D. Because there is less to eat the predator population then also decreases.

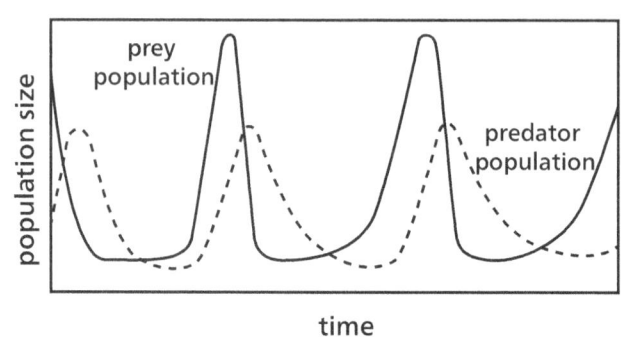

217

 **BIOLOGY HIGHER**

Main activity: Cycling of materials and decomposition   Time **15** mins

Learning objectives

- To explain the importance of the carbon and water cycles to living organisms
- To explain how temperature, water and availability of oxygen affect the rate of decay of biological material

Equipment

none

1. All materials in the living world must be recycled to provide the building blocks for future organisms, such as in the carbon cycle. These materials cycle through the abiotic and biotic components of an ecosystem. The diagram below shows some of the processes involved in the carbon cycle.

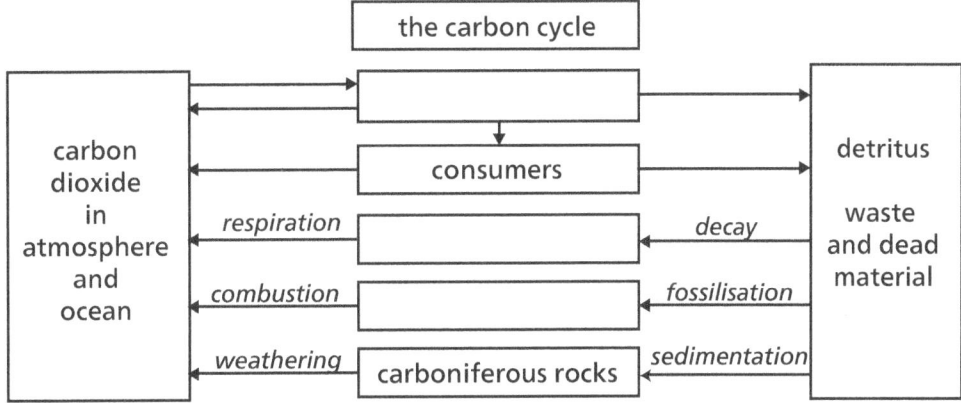

   a) Write the following labels into the correct blank box: fossil fuels   producers   microorganisms

   b) Write these labels onto arrows wherever it is appropriate: respiration   photosynthesis   death   feeding

   c) Microorganisms are very important in cycling materials through an ecosystem. By the process of decomposition (decay) they return carbon to the atmosphere as carbon dioxide. Why is this important for plants?

   d) How does decomposition by microorganisms help to improve the soil for plants?

2. Microorganisms cause milk to decay. They produce enzymes which break down the milk and produce acids. Salma is planning to investigate the effect of temperature on the rate of decay of milk by measuring pH change. She will use lipase enzyme to model the enzymes of the bacteria. Lipase breaks down fat in milk to make fatty acids. Write some notes in the table below to help Salma plan the investigation. Include any important equipment.

Independent variable – changing the temperature	Dependent variable – measuring rate of pH change
Controlled variables – the same at all temperatures	Safety notes

218

BIOLOGY HIGHER

Homework activity: Decay and sampling

Time 30 mins

Learning objectives
- To explain the optimum conditions for decay
- To describe how biogas can be produced

Equipment
- calculator

1. Waste organic material, such as manure or plant waste, can be used to make biogas fuel. A simple diagram of a biogas generator is shown below.

   a) Oxygen must not be introduced into the waste tank. Explain why this is important.

   _____
   _____
   _____

   b) What is the name of the gas that is produced by this type of decay?
   _____

   c) Some biogas generators have solar panels that heat up sewage. Explain why this might be useful.
   _____

   d) It is important that the temperature in the biogas generator does not rise too high. Explain why.
   _____

2. Gardeners and farmers put waste biological material on a compost heap and leave it to decay.

   a) What would be the best conditions for decay of the biological waste to make compost? Describe three conditions that would give rapid decay.
   _____

   b) What is the compost used for?
   _____

3. Daisies were counted in ten 0.25 m^2 quadrats from a field that is 500 m^2. Results (number per quadrat) are shown below.

25	12	23	41	43	32	18	24	19	28

Calculate the following:

Mean: _____   Median: _____

Use your mean to estimate the daisy population of the field: _____

34 Answers

Starter activity: Food in a chain, water in a cycle

1. Chains, producer, alga, glucose, primary, secondary, tertiary, predators
2. Grass → rabbit → fox
3. Annotations should show that the water is continuously evaporating and precipitating.

Main activity: Populations, predators and prey

1. a) See glossary on page 245
   b) It would be too time consuming to count all the plants, and some might be missed.
   c) Place the quadrats randomly in each area/use random numbers on a grid; count the number of plants lying in each quadrat, work out the number per m^2, find a mean; to overcome edge effect count all plants that cross two of the sides (even by a small amount) and no plants that cross the two other sides
   d) Use percentage cover to estimate the percentage of ground covered by the plant in each quadrat. Use a squared quadrat or point quadrat to help.
   e) Use a transect. Lay a tape measure along a line from the shaded area to the open area and place quadrats at even intervals along the line. Record abundance in each quadrat and plot abundance against distance along the transect.
   f) Light intensity, with a light meter, at ground level, as close in time as possible and multiple measurements at each site
2. A at any point where prey numbers are rising; B the following section of rising predator numbers; C at any point where prey numbers are declining; D at any point where predator numbers are declining

Main activity: Cycling of materials and decomposition

1. a) Order going down boxes: producers, microorganisms, fossil fuels
   b) Respiration on top two arrows pointing left; photosynthesis on arrow pointing to producers; death on top two arrows pointing right; feeding from producers to consumers
   c) Needed for photosynthesis
   d) They return mineral ions to the soil.
2. Temperature: at least five even intervals, suitable range, such as 20–60 °C, allow time to reach correct temperature, waterbath, thermometer; pH change: use indicator/creosol red, time how long until it changes colour for acid/turns yellow, stop clock; controlled: concentration of lipase enzyme, volume of milk and indicator, starting pH; safety notes: take precautions to avoid scalding, safety glasses, avoid skin contact with enzyme and creosol red

Homework activity: Decay and sampling

1. a) The decay must be anaerobic; otherwise fuel gas would not be produced/CO_2 would be produced if O_2 present
   b) methane
   c) Decay/gas production/anaerobic respiration would be faster in warm conditions
   d) At high temperatures, the microorganisms responsible for decay would be killed/proteins would be denatured
2. a) Oxygenated/aerobic conditions; warm temperatures; moist conditions
   b) The compost produced is used as a natural fertiliser for growing garden plants or crops.
3. Mean 27 (26.5); median 24.5; population 53 000 (26.5/0.25=106 m^{-2}, 106 × 500 = 53 000)

35 Ecology: Environmental change, biodiversity and waste management

Learning objectives	Specification links
• To understand how environmental change can affect species distribution	• 4.7.2.4
• To understand the importance of biodiversity	• 4.7.3.1
• To understand how the human population can cause pollution of water, air and land	• 4.7.3.2

Starter activity

- **Environmental change; 10 minutes; page 222**

  The student should focus on how environmental change influences species distribution. Changes might be seasonal, geographic or caused by human interaction. They may find it easier to think of specific examples (such as temperature and polar bears) rather than general points.

Main activities

- **Species distribution and biodiversity; 20 minutes; page 223**

  The first part is a data interpretation exercise based on the change in composition of atmospheric gases. Questions could be answered verbally or on a separate piece of paper. Question 2b) should be answered on the activity sheet, while part c) should form the basis of a discussion.

- **Waste management; 15 minutes; page 224**

  Questions 1 and 2 should be answered on the sheet and questions 3–5 answered verbally. The student should be given time to record their own ideas for question 2, but may need some prompts initially. Check that they are familiar with at least the most important facts about each type of pollution.

Plenary activity

- **One-minute summary; 5 minutes**

  First, briefly review each part of the lesson and the links between them. Then challenge the student to talk fluidly for one minute without hesitating on the topic of 'humans and biodiversity'.

Homework activity

- **Water pollution; 20 minutes; page 225**

  This homework consists of two exam-style questions on water pollution, fertiliser pollution and the accumulation of chemicals in food chains.

Support ideas

- **Species distribution and biodiversity** Show pictures to help the student to visualise the points about environmental change and distribution, or to understand what a lichen is. An internet search for 'climate change and desertification map' will give useful results to demonstrate how desert ecosystems may spread in future years.
- **Waste management** Show graphs to help the student to visualise the pattern of human population growth over the last few hundred years. Ask them to think about what they own compared with what their grandparents might have owned and discuss the materials used in modern possessions such as TVs, phones and cars.

Extension ideas

- **Species distribution and biodiversity** Ask the student to explain what would happen to the whole ecosystem in each case if biodiversity was low.
- **Waste management** Ask the student to explain how some chemicals accumulate in food chains and what the consequence of this might be.

Progress and observations

BIOLOGY HIGHER

Starter activity: Environmental change Time 10 mins

Learning objectives

- To understand how environmental changes can affect the distribution of species in an ecosystem

Equipment

none

1. Environmental changes can affect the distribution of species in an ecosystem. Three factors that can change and may affect where species are found are shown in the table below. For each factor write down some examples of how it may affect the distribution of organisms. One example for each has been given for you.

Environmental factor	How changes may affect the distribution of organisms
Temperature	An increase in temperature may increase growth rates for some plants, changing competition and the balance of species.
Water availability	An increase in rainfall may cause waterlogged soil, reducing soil air content so that aerobic bacteria are replaced by anaerobic ones.
Composition of atmospheric gases	Sulfur dioxide from the burning of fossil fuels dissolves to make acid rain. This reduces plant growth and reduces populations of some plant species.

Main activity: Species distribution and biodiversity Time 20 mins

Learning objectives

- To evaluate the impact of environmental changes on the distribution of species in an ecosystem
- To understand the importance of biodiversity

Equipment

none

1. The composition of gases in city areas in the UK has shown a number of changes in the latter half of the twentieth century. While some air pollutants increased, others decreased. One air pollutant that decreased during this time is sulfur dioxide, which is released by the burning of fossil fuels, such as coal. Some species of lichen are very sensitive to sulfur dioxide. The table shows some changes in lichens seen at some urban sample sites.

Year	Percentage of sites with crusty lichens	Percentage of sites with leafy lichens	Number of lichen species recorded
1950	92	0	1
1960	94	0	1
1970	91	0	2
1980	83	5	4
1990	45	22	6
2000	35	43	5
2010	28	46	6

a) Lichens are used as bioindicator species for sulfur dioxide pollution. What does this mean?

b) Sulfur dioxide concentrations decreased over the period of study. Which type of lichen is most tolerant of sulfur dioxide pollution?

c) Suggest why the number of species of lichen recorded increased over the time period that was studied.

d) Sulfur dioxide dissolves in water and causes acid rain. Describe how acid rain can affect two other species.

2. **Biodiversity is the variety of all the different species of organisms on Earth or within an ecosystem.**

a) Many human activities are reducing biodiversity. List as many of these human activities as you can.

b) High biodiversity keeps ecosystems stable. Some reasons for this are given below. Draw lines to link each reason for improved ecosystem stability to the relevant example.

Good biodiversity reduces the dependence of one species on another for food.	There is a positive correlation between the diversity of coral reef fish and the variety of three-dimensional coral structures available. Different coral species have different shapes.
Good biodiversity provides alternatives for shelter.	A plant species within a community suffers due to the appearance of a new disease. A herbivore that usually feeds on this plant begins to eat a different species instead.
Good biodiversity helps to maintain the physical environment.	In a diverse woodland community, the roots of each plant species help to bind soil together and reduce soil erosion.

c) Explain to your tutor why the future of the human species on Earth relies on us maintaining a good level of biodiversity.

BIOLOGY HIGHER

Main activity: Waste management   Time 15 mins

Learning objectives
- To understand how the human population can cause pollution of water, air and land

Equipment
none

1. The global human population has increased rapidly in recent centuries and more than doubled between the 1960s and the end of the last century. This increase puts pressure on the world's resources and increases the amount of waste that is produced. What other change in human population has also used up more natural resources and produced more waste?

2. Waste and chemical materials, unless properly handled, can cause pollution of air, land and water. For each of the pollutants below, add details to the table to explain what effects they have on ecosystems.

Air pollution	Water pollution	Land pollution
smoke	sewage	landfill
	fertiliser	
acidic gases		toxic chemicals
	toxic chemicals	

3. Toxic chemicals from farmland include pesticides and herbicides. Explain to your tutor why these chemicals are used.

4. Pollutants in the air may affect areas far away. For example, radioactive pollution from the Chernobyl nuclear power station disaster caused pollution of hill land in the UK. This meant that sheep reared there could not be eaten for many years. How can this type of pollution travel so far?

5. How are many governments now trying to reduce the amount of new landfill sites that are needed?

BIOLOGY HIGHER

Homework activity: Water pollution   Time 20 mins

Learning objectives
- To explain the effects of water being polluted with fertiliser

Equipment
none

1. Fertilisers are chemicals that are used to improve crop growth and increase yields. They can be washed off farmland and into rivers where they cause a type of pollution known as eutrophication. Some of the steps that take place in eutrophication are given in the table below. They are not in the correct order.

A.	The minerals in the fertiliser make the plant plankton and other water plants grow rapidly.
B.	The numbers of microorganisms increase rapidly.
C.	These microorganisms use up a lot of oxygen, the oxygen levels dissolved in the water become very low.
D.	Competition for light increases greatly.
E.	Biodiversity in the water body becomes very low.
F.	The fertiliser dissolves in rainwater and runs into streams, lakes or rivers.
G.	Plants begin to die.
H.	Fish and other animals living in the water die.
I.	Microorganisms start to feed on the dead plants and cause them to decay.

Write down the letters that show the correct order of the processes that lead to eutrophication. The first one has been done for you.

F								

2. Some chemical pollutants build up in plants and animals, and are found at higher levels towards the top of the food chain. This can happen on land or in water. One example of this is shown below for an aquatic community.

	Concentration of industrial chemical 'X' (ppm)
Water	0.0001
Producers	1
Herbivores	10
Carnivores	50

Explain how some chemicals become concentrated at higher levels further up the food chain.

_____
_____
_____
_____
_____
_____

35 Answers

Starter activity: Environmental change

1. There are many possible responses for this activity which will depend on the student's knowledge. Answers should focus on distribution of species; ideas might include: distribution of cold-adapted organisms become restricted further towards poles or further up mountains; extinction; decrease in marsh communities or spread of desert communities; decreasing water availability; changes in oxygen content of atmosphere over billions of years due to photosynthesis caused major changes in species on Earth; acidic gases, such as sulfur dioxide, cause acid rain; some lichen species more tolerant, some species die off

Main activity: Species distribution and biodiversity

1. a) Their presence or absence can be used to assess the amount of a certain type of pollution.
   b) Crusty lichens
   c) Only one species was adapted to survive in acidic conditions. When SO_2 levels reduced more species could survive. With more competition, the distribution of the tolerant species declined.
   d) Reduces tree growth, damages roots and leaves of trees, leaches aluminium from soil into water courses which kills aquatic life
2. a) Some examples include: pollution; human population increase; destroying habitats; deforestation; drainage of wetlands; being eaten for food; planting crops; flooding valleys for hydro schemes
   b) Food: plant species; shelter: coral reef fish, physical environment: woodland roots
   c) There are many possible examples. Answers should rely on the ecosystem services that high biodiversity can provide, for example: reservoir of genetic diversity which may help provide disease-resistant crop and animal varieties; undiscovered medicines from wild plant species; stable ecosystems help to regulate climate and atmospheric composition; possible new food sources; many insects are pollinators of crop plants

Main activity: Waste management

1. The increase in the standard of living, as more materials are used per head
2. Smoke: global dimming, reduces photosynthesis, food chain effects; acidic gases: nitrogen oxides and sulfur dioxide, kills species, reduces biodiversity, acid rain; sewage: deoxygenation of water, smothering, reduces biodiversity; fertiliser: eutrophication, algal blooms; landfill: methane (greenhouse gas) released, leaching waste chemicals, takes up space (habitat loss); toxic chemicals: poisons kill species and reduce biodiversity, may accumulate in food chains, for example heavy metals and PCBs, may also cause water pollution – toxic chemicals in water often include pesticides from farmland
3. To kill insect or other pests that eat the crop; to kill weeds that compete with the crop and reduce yields
4. Carried on the wind (and in water/rain)
5. Encouraging reuse, recycling and a reduction in packaging

Homework activity: Water pollution

1. F, A, D, G, I, B, C, H, E
2. Any four points from: plants absorb the chemical at low levels from the water in which they live; it builds up to higher levels in their tissues than in the water; herbivores eat lots of plants that have high concentrations of the chemical; they cannot get rid of/break down the chemical, the chemical becomes more concentrated in herbivores; carnivores receive even higher doses when they eat the herbivores

36 Ecology: Land use, deforestation and global warming

Learning objectives

- To understand how land use can affect ecosystems and biodiversity
- To describe some of the biological consequences of global warming
- To describe programmes that have been put in place to reduce negative effects

Specification links

- 4.7.3.3
- 4.7.3.4
- 4.7.3.5
- 4.7.3.6

Starter activity

- **Land use; 5 minutes; page 228**

  The student should complete the activities independently. Check their answers and then, if there is time, ask the student to reword the false answers so they become true.

Main activities

- **Peat bogs and global warming; 20 minutes; page 229**

  Provide sufficient time for the student to read the information on peat bogs then ask them to complete the questions by circling correct answers. There is more than one correct answer in each case. Emphasise the importance of microorganisms. Question 7 is best answered with the table provided as cut out cards to sort, but it can be done by drawing lines.

- **Deforestation and maintaining biodiversity; 15 minutes; page 230**

  The first question focuses on deforestation, which is picked up again in the homework. The student may need help to fill in the table. In the second question, they should add to the diagram using their own background knowledge where appropriate.

- **The causes of global warming; 5 minutes**

  Review the causes of global warning focusing on how certain gases in the atmosphere trap heat radiated from the Earth – ask the student to explain why it is called the 'greenhouse effect'. Ask them to list sources of methane and CO_2.

Plenary activity

- **Note to tutor; 5 minutes**

  Ask the student to write down the most important thing they learned in the lesson and one question that they still wish to ask.

Homework activity

- **The impacts of deforestation; 15 minutes; page 231**

  The homework is an extended answer question covering the environmental effects of deforestation.

Support ideas

- **Peat bogs and global warming** Many students will never have seen a peat bog or sphagnum moss. Show the student photos of peat bog habitats, peat bricks drying for burning and extracted compost.
- **Deforestation and maintaining biodiversity** Show maps or graphs to help the student's understanding of the scale and location of deforestation. An internet search for 'global deforestation map' will give some useful results.

Extension ideas

- **Peat bogs and global warming** Ask the student if they would consider peat to be a renewable fuel and to justify their answer. This is a controversial question as the peat industry has stated that peat is a 'slowly renewable' fuel because it takes tens of thousands of years (rather than millions for fossil fuels) to be replaced naturally.
- **Deforestation and maintaining biodiversity** Ask the student to take the position of a government in a developing country that has a large forest resource. Ask them to state the argument for cutting down the forest (for example, land for food crops, sale of wood, jobs, economic growth and to provide biofuel).

Progress and observations

BIOLOGY HIGHER

Starter activity: Land use

Time **5** mins

Learning objectives
- To understand how land use can affect ecosystems and biodiversity

Equipment
none

1. Look at the symbols in the table and use them to complete the following statement.

Humans reduce the amount of land available for other animals and plants by:

house, saw	B...
mountains, excavator	Q...
tractor, cow	F...
bin with waste	D................... W...................

2. For each of the following statements circle T if you think it is true and F if you think it is false.

a) Peat is a fuel that is made from felled trees.   T F

b) Burning peat releases CO_2 into the atmosphere.   T F

c) Carbon dioxide is the most powerful greenhouse gas.   T F

d) Photosynthesis in forests may help to reduce global warming.   T F

e) Biofuel is fuel produced from crops or biological waste.   T F

f) Biofuels could replace the global use of fossil fuels.   T F

g) Methane is a greenhouse gas that is produced by anaerobic decay.   T F

BIOLOGY HIGHER

Main activity: Peat bogs and global warming   Time 20 mins

Learning objectives

- To understand the conflict between the need for compost and the need to conserve peat bogs and peatlands
- To describe some of the biological consequences of global warming

Equipment

- cards to sort for question 2 (optional)

1. Read the information about peat bogs, then underline the correct answers to the questions.

> Peat bogs cover almost 3% of the Earth's surface and are an important carbon sink. These habitats support unique ecosystems with diverse plant, animal and microorganism species. Peat is a material that is formed when moss grows on waterlogged soils. Waterlogged areas provide anaerobic conditions which, together with an acidic pH, prevent the moss from completely decomposing when it dies. As the mosses photosynthesise, they absorb carbon dioxide. When the moss dies, this carbon is not returned to the atmosphere by microbial decay, but is locked away in the peat. It takes thousands of years for peat to build up to a few feet depth.
> Peat can be dried and burnt as a fuel. Peat is also a source of cheap, readily available compost that can be added to soil to increase crop production. When mixed into soil, it improves soil structure, mineral availability, water retention and acidity. Many peat bogs have been drained to allow the peat to be dug out. Burning peat, draining peatlands and mixing extracted peat into soil as a compost all return carbon dioxide back into the atmosphere.

a) What conditions are needed for peat to form?

soil is: well aerated / little oxygen in soil / dry / wet / acidic / alkaline

b) Why do these conditions cause peat to form?

aerobic decay is reduced / microorganisms grow best in moist conditions / it is too acidic for most microorganisms

c) Why is peat a popular choice as a compost?

cheap / reduces acidity of soil / improves soil quality / drains quickly / widely available

d) Why does draining peat areas and using peat as a compost contribute to global warming?

reduces anaerobic decay / increases aerobic decay / peat only forms in waterlogged areas / releases trapped CO_2

e) How can the two main problems of using peat be summarised?

loss of biodiversity / expensive / contributes to global warming / needs transporting / plants may dislike acid soil

f) What measures could be used to protect peat habitats and reduce the impact on global warming?

use compost made from green waste / government targets to protect peat / ban the use of compost

2. Global warming may affect biological systems in many ways. Match up each change to the possible consequences.

Change	Consequence for organisms
new temperature may be too high for the species to thrive or complete development	by increasing photosynthesis
warmer climate may suit a different species	species may move to cooler regions instead
increasing CO_2 may benefit plants	species abundance may change
changes in biotic factors affecting a species such as: food availability, predation, competition or disease	native species are outcompeted by invasive species
if climate change is slow	species are likely to become extinct
if climate changes very rapidly	animals may emerge or hatch at the wrong time and there may not be enough available food
warmer temperatures may affect life cycles	species may evolve and adapt

BIOLOGY HIGHER

Main activity: Deforestation and maintaining biodiversity

Time 15 mins

Learning objectives

- To understand why large-scale deforestation has occurred
- To describe programmes to reduce the negative effects of human activity on ecosystems and biodiversity

Equipment

none

1. Large-scale deforestation in tropical areas has occurred over the last century. The conservation body WWF reports that an area the size of England is lost every year.

   a) What are the two main problems linked to the loss of tropical forests? Discuss each one with your tutor.

   _____ and _____

   b) The table lists three of the reasons why large areas of forest have been cut down. For each one, describe any problems specific to this land use.

To provide land for cattle	To provide land for rice fields	To grow crops for biofuels

2. Scientists and concerned citizens have put in place programmes to reduce some of the negative effects of humans on biodiversity. Some of these programmes are outlined below. Add at least one example of each to the diagram.

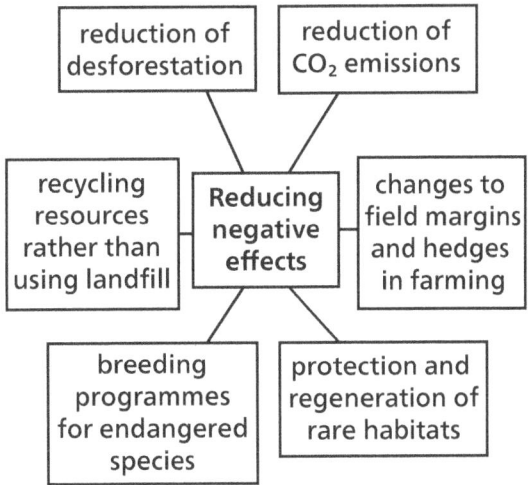

BIOLOGY HIGHER

Homework activity: The impacts of deforestation

Time 15 mins

Learning objectives
- To evaluate the environmental implications of deforestation

Equipment
none

1. Answer this exam-style question.

> Deforestation has taken place in many tropical areas of the world. In many areas, tropical rainforest has been replaced with crops such as palm oil trees, which are used to make biofuel.
>
> Evaluate the environmental effects of deforestation to produce biofuel crops. You should include both the environmental problems and any environmental benefits.
>
> [6 marks]

BIOLOGY HIGHER

36 Answers

Starter activity: Land use
1. Building, quarrying, farming, dumping waste
2. a) F; b) T; c) F; d) T; e) T; f) F; g) T

Main activity: Peat bogs and global warming
1. a) Little oxygen in soil/wet/acidic
   b) Aerobic decay is reduced/it is too acidic for most microorganisms
   c) Cheap/improves soil quality/widely available
   d) Increases aerobic decay/releases trapped CO_2
   e) Loss of biodiversity/contributes to global warming
   f) Use compost made from green waste/government targets to protect peat

2.
New temperature may be too high for the species	Species may move to cooler regions instead
Warmer climate may suit a different species	Native species are outcompeted by 'invasive' species
Increasing CO_2 may benefit plants	By increasing photosynthesis
Change in biotic factors affecting a species such as: food availability, predation, competition or disease	Species abundance may change
If climate change is slow	Species may evolve and adapt
If climate changes very rapidly	Species are likely to become extinct
Warmer temperatures may affect life cycles	Animals may emerge or hatch at the wrong time and there may not be enough available food

Main activity: Deforestation and maintaining biodiversity
1. a) Loss of biodiversity/extinctions; increased atmospheric CO_2 concentrations
   b) Cattle produce methane, a more powerful greenhouse gas than carbon dioxide, contributing to global warming. Flooded rice fields encourage microorganisms, which cause anaerobic decay, which produces methane as organic material rots. Biofuels are grown on land which cannot then be used to produce food, and need lots of fertiliser and water for irrigation.
2. Fields: leave a wild strip around field/plant hedge; creates more biodiversity where usually a monoculture; breeding programmes in zoos and reserves: many species such as pandas; deforestation and CO_2 emissions: some governments set targets/sign international agreements (for example the Paris Agreement has both deforestation and emissions targets); recycling: paper, green waste, glass, metal and so on; habitat protection: nature reserves, regeneration

Homework activity: The impacts of deforestation

The following table provides guidance on what a Level 3, 2 or 1 answer to this question would look like and the number of marks each would attract.

L3	A detailed and coherent evaluation is provided that considers a range of relevant points, including, balanced argument, and comes to a conclusion consistent with the reasoning.	5–6 marks
L2	Attempts to describe relevant points and comes to a conclusion. May lack balance but is coherent.	3–4 marks
L1	Some, relevant points made. The organisation may be unclear and a conclusion may not be made.	1–2 marks

Indicative content	
Environmental problems: • Deforestation contributes to global warming • Increase in carbon dioxide in atmosphere • Due to burning of forest trees • Due to decay of felled trees by of microbes • Biofuel crop has less biomass/palm trees smaller than rainforest community/trees • Less carbon dioxide taken in/locked up by trees • Less photosynthesis	• Forest habitats are destroyed and species become extinct • Biodiversity is reduced *Environmental benefits:* • Biofuel can replace/reduce CO_2 released from burning fossil fuels • Used for biodiesel/transport fuel • Biofuel may be regarded as carbon neutral

BIOLOGY HIGHER

37 Ecology: Trophic levels in an ecosystem

Learning objectives

- To describe differences between the trophic levels in an ecosystem
- To construct accurate pyramids of biomass
- To explain and calculate biomass loss between different trophic levels

Specification links

- 4.7.4.1
- 4.7.4.2
- 4.7.4.3
- MS 1c, 2c

Starter activity

- **Trophic levels; 5 minutes; page 234**

  Ask the student to complete this activity independently. It should be straightforward for most higher level students.

Main activities

- **Pyramids of biomass; 20 minutes; page 235**

  Question 2 should be answered through a verbal discussion. Question 5 should be done on graph paper. The student should be able to plot a pyramid of biomass to scale and label the levels. Point out that these numbers are unusual, as normally the biomass of the producer would be relatively larger, and that this can be difficult to plot.

- **Transfer of biomass; 15 minutes; page 236**

  Support the student in question 2 by drawing a larger diagram on a separate piece of paper if necessary. Question 3 reinforces the idea of biomass loss via respiration. Question 4 covers biomass efficiency calculations for food chains. Emphasise that this is an oversimplification; not all of the biomass from one level would be eaten by the next. Numbers on the arrows may be used to show energy transfer, or how much biomass is actually eaten by the next level.

- **Energy intake and losses; 5 minutes**

  Ask the student to represent the biomass gains and losses by an organism as an equation, showing transfer into the biomass (B) of the organism, in order to complete B = ? Use the symbols I (ingested), R (respiration), F (faeces) and U (urine). When answered (B = I – R - F – U), ask them to rearrange it to show ingested biomass (giving: I = B + R + F + U).

Plenary activity

- **Word tennis; 5 minutes**

  Play word tennis. Say a keyword from the lesson, and the student must respond with a linked word. A person scores when the other person hesitates. Use tennis scoring, and a real ball can be passed back and forth. Review the meaning of any words the student is uncertain about.

Homework activity

- **Pyramids and energy loss; 25 minutes; page 237**

  The homework consists of short, exam-style questions including plotting a pyramid of biomass.

Support ideas

- **Trophic levels** If the student is initially unsure about the differences between trophic levels, explain, then ask them to add the levels and descriptions of feeding relationships to their sketch of the oak wood pyramid of biomass.
- **Transfer of biomass** The student may need more practice at calculating percentage efficiencies of biomass transfer. If so, use the marine food chain on the starter page, or make up some simple numbers for other food chains.

Extension ideas

- **Pyramids of biomass** Discuss how decomposers might be added to pyramids of biomass. Show an example where decomposers are added as a vertical bar to the side of the pyramid. Ask the student to sketch this for the oak wood example.
- **Transfer of biomass** Ask the student to explain the difference between the terms egested and excreted, and to explain where urea comes from in metabolic terms. They should recall that it is a waste product made from excess amino acids.

Progress and observations

Starter activity: Trophic levels

Time 5 mins

Learning objectives
- To describe differences between the trophic levels in an ecosystem

Equipment
- pencil
- spare parper

1. The term 'trophic levels' refers to the different feeding levels in an ecosystem. Trophic levels can be represented by numbers, starting at level 1. Draw lines to connect the trophic level on the left with the relevant type of organism in the middle and the descriptive term on the right.

level 1	herbivores (eat plants/algae)	primary consumers
level 2	plants and algae that make their own food	secondary consumers
level 3	carnivores that eat other carnivores	tertiary consumers
level 4	carnivores that eat herbivores	producers

2. A simple food chain found in an oak wood might be as follows:

oak tree → caterpillar → vole → tawny owl

On a separate piece of paper, for this food chain do a rough sketch of:

a) a pyramid of numbers

b) a pyramid of biomass

c) Which of these would be the most useful to describe energy flow within the ecosystem?

_____
_____

BIOLOGY HIGHER

Main activity: Pyramids of biomass

Time 20 mins

Learning objectives
- To understand the concept of trophic levels in an ecosystem
- To construct accurate pyramids of biomass

Equipment
- graph paper
- pencil

1. Pyramids of biomass are used to show feeding relationships and represent the relative amount of biomass in each level of a food chain. Add labels for the trophic levels to the pyramid of biomass below.

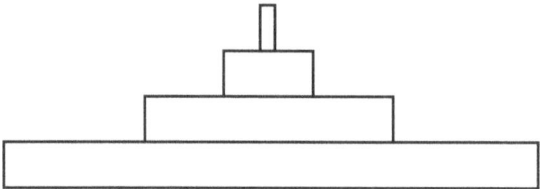

2. Discuss the following with your tutor:

    a) What is meant by the term apex predator?

    b) What organism would this be in the food chain in the starter activity?

    c) Where does all the energy in living organisms in food chains on land come from?

    d) How does this compare with marine food chains?

3. Fill in the gaps to complete the following sentences:

    Decomposers break down _____ plant and animal matter by secreting _____ into the environment. Small _____ food molecules then _____ into the microorganism.

4. Why are decomposers not included within pyramids like the one shown above? _____

5. Biomass is the mass of all living material in an area. It can be expressed as total mass or mass per unit area, but is usually given as a mass of dry material. Why? How would this be determined?

6. On graph paper, plot out the pyramid of biomass for the following marine food chain.

Trophic level	Organism	Total Biomass (kg)
1	algae	80
2	limpet	20
3	starfish	2

7. Pyramids of biomass are usually pyramid shaped. Occasionally they are not. For example, marine food chains based on plant plankton may look like this at some times of year. Can you explain why?

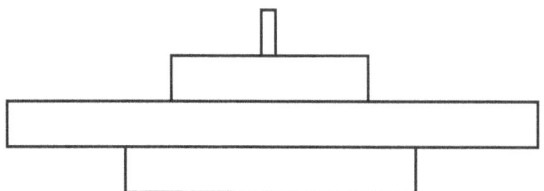

BIOLOGY HIGHER

Main activity: Transfer of biomass   Time 15 mins

Learning objectives
- To explain how biomass is lost between the different trophic levels
- To calculate efficiency of biomass transfer between trophic levels

Equipment
- pencil
- calculator

1. Producers are mostly plants and algae which transfer about 1% of the incident energy from light for photosynthesis. As this energy moves up the food chain, only approximately 10% of the energy in biomass from each trophic level is transferred to the level above it. Show this as a flow diagram following 1000 J of energy from sunlight through a food chain.

sunlight	→	producers	→	primary consumers	→	secondary consumers
1000 J		J		J		J

2. Energy is locked up in biomass, so there are large losses of biomass at each stage. Any biomass that is not lost will become part of the biomass of the organism. Add arrows and labels to the diagram below to show what happens to the biomass in the carrot. Make sure that you use the following words: ingested, egested, faeces, absorbed, biomass, respiration, urine.

3. Respiration is the main mechanism by which biomass (and energy) from food is lost and is not transferred into the biomass of an organism.

   a) What substance enters as a reactant in respiration in large amounts? _____

   b) How is this biomass then lost to the atmosphere? _____

4. We can compare the efficiency of biomass transfers between trophic levels using percentages or fractions of mass. The biomass in g m^{-2} of a field is shown for a simple food chain.

grass	→	crickets	→	blackbirds
1029		83		8

   a) What is the percentage of biomass in the grass that is transferred into biomass of the crickets? _____

   b) Approximately what fraction of the biomass from crickets ends up as biomass in blackbirds? _____

   c) What percentage of the biomass from the producer in this food chain is transferred into biomass of the secondary consumer?

   _____

BIOLOGY HIGHER

Homework activity: Pyramids and energy loss

Time 25 mins

Learning objectives
- To construct accurate pyramids of biomass
- To calculate efficiency of biomass transfer between trophic levels

Equipment
- pencil
- ruler
- calculator

1. **The table shows the biomass of organisms in a food chain in a pond.**

   a) On the grid below, draw a pyramid of biomass for this food chain. Draw the pyramid to scale.

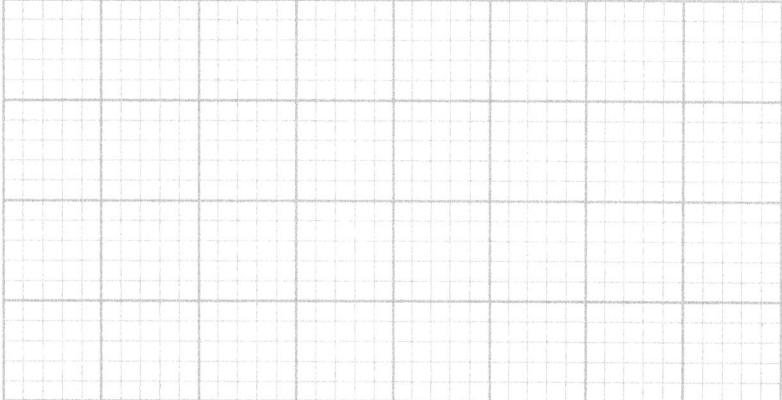

Organism	Biomass (g m^{-2})
water plants	80
freshwater shrimps	30
dragonfly nymphs	6
trout	1

   b) Calculate the percentage of the biomass of the water plants that is transferred into the biomass of the shrimps.
   _____

   c) What is the ratio of the biomass of the freshwater shrimps to that of the dragonfly nymphs?
   _____

   d) It is very unusual for there to be more than five trophic levels in any food chain. Explain why.
   _____
   _____
   _____

   e) Bacteria that are decomposers are an important part of the pond ecosystem. Explain how decomposers feed.
   _____
   _____

37 Answers

Starter activity: Trophic levels

1. Level 1, plants and algae, producers; level 2, herbivores, primary consumers; level 3, carnivores that eat herbivores, secondary consumers; level 4, carnivores that eat other carnivores, tertiary consumers
2. a) Four levels, the bottom level (oak) should be narrower than the one above, then decreasing width
   b) Four levels, pyramid shaped
   c) Biomass is more useful to describe energy flow.

Main activity: Pyramids of biomass

1. Trophic level 1 at the bottom of the pyramid, four levels labelled
2. a) Apex predators: carnivores with no predators
   b) Tawny owl
   c) The Sun
   d) Marine ecosystem: most rely on the Sun, some deep sea ecosystems rely on chemical energy from deep sea vents
3. Decomposers break down dead plant and animal matter by secreting enzymes into the environment. Small soluble food molecules then diffuse into the microorganism.
4. Because they feed on dead organisms from every level of the pyramid
5. Dry mass is more representative of energy content; wet weight can be variable, especially for plants; dry to a constant mass in a warm oven (not hot enough to burn)
6. Pyramid with algae on the bottom level, all levels to scale and centred around a central line
7. Biomass is given at one point in time only. Where there are seasonal changes in abundance, the producer may have been eaten and its biomass transferred into the next trophic level.

Main activity: Transfer of biomass

1. Producer 10 J, primary consumer 1 J, secondary consumer 0.1 J
2. Labels should cover the following points: carrot biomass is ingested; some is not absorbed - egested as faeces; rest of material is absorbed; some of absorbed material lost from respiration (CO_2 and water); some absorbed material is lost as water and urea in urine
3. a) glucose
   b) Carbon dioxide and water
4. a) 8.1 %
   b) 1/10
   c) 0.8%

Homework activity: Pyramids and energy loss

1. a) All four levels drawn to scale (trout can be a line); all levels plotted evenly around a central line; each level labelled
   b) 37.5%
   c) 5 : 1
   d) A lot of biomass is lost at each trophic level; example of how energy is lost (in faeces, respiration/CO_2 and water, in urine/water and urea); there is not enough biomass left at the top of the food chain to provide enough food for another level
   e) They break down dead plant and animal matter; secrete enzymes; absorb the small/soluble food molecules produced

BIOLOGY HIGHER

38 Ecology: Food production

Learning objectives

- To describe some of the biological factors affecting levels of food security
- To describe some methods that can improve the efficiency of farming and the sustainability of fishing
- To explain some biotechnical solutions to maintain food security

Specification links

- 4.7.5.1
- 4.7.5.2
- 4.7.5.3
- 4.7.5.4

Starter activity

- **Food security; 10 minutes; page 240**

  Provide the student with sort cards, not the whole activity sheet. They should first match up the factors and consequences, then rank the factors in terms of relative effect on food security. There is no correct answer to this, but it should prompt discussion at the end of the activity. Population growth and environmental changes due to climate change are probably the most significant for future global food supply.

Main activities

- **Farming and fishing techniques; 15 minutes; page 241**

  These activities cover farming methods and fishing sustainably. These should be answered verbally as a discussion, but the student should be encouraged to note down key points.

- **The role of biotechnology; 15 minutes; page 242**

  Ask the student to complete each part of the activity and then discuss their answers. Some explanation of the fermenter may be needed. The risks and benefits of GM crops and insulin production were also considered in lesson 29.

- **Golden rice; 5 minutes**

  Ask the student to argue the case for golden rice playing a role in food security. Golden rice is a variety produced through genetic engineering that has genes for synthesis of a precursor of vitamin A. Vitamin A deficiency is a common form of malnutrition in the developing world, causing a variety of problems.

Plenary activity

- **One fact, one question; 5 minutes**

  Using the six 'biological factors affecting food security' cards from the starter activity, ask the student to choose two of these cards. For each chosen card, they must then tell you one fact and ask you one question.

Homework activity

- **Farming efficiency; 15 minutes; page 243**

  The homework activity consists of short exam-style questions focused on the efficiency of food production

Support ideas

- **Food security** Provide examples linked to the causes of food security. Search online to find information on the causes behind particular famine events of the past such as in Ireland 1845–52, Somalia, Zimbabwe, Ethiopia or Cambodia.
- **The role of biotechnology** It may be necessary to revisit the meaning of genetic modification. Discuss the addition of useful genes. Ask for suggestions about the origin of these genes.

Extension ideas

- **Farming and fishing techniques** Describe intensive fish farming where fish are kept in cages and fed on pellets that are made from smaller fish species. Ask the student to suggest the benefits (efficient biomass transfer, reduced movement) and problems (pollution from waste and pesticides) of fish farming.
- **The role of biotechnology** Ask the student to suggest what restrictions there may be on the impact of biotechnology on food security. Key is populations that most need help are least likely to be able to afford or access the technology.

Progress and observations

Starter activity: Food security

Time 10 mins

Learning objectives
- To describe some of the biological factors affecting levels of food security

Equipment
cut out set of the boxes below to make sort cards

1. Food security means having enough food to feed a population. Globally, there are many biological factors which threaten food security. Your tutor will provide you with cut outs of the boxes below.

   Match each factor (in bold) to its consequence for food production. When you have finished, think about the relative importance of each factor in terms of impact on food security, then put the cards in order from most to least important.

increasing human birth rate in some countries	There may not be enough food to feed the growing population. There may not be enough suitable land available to grow the food needed.
changing diets in developed countries	Scarce food resources are transported around the world. Countries where food resources are low may grow crops to export for profit, rather than grow crops to feed local people.
new pests and pathogens	These may destroy crops on a large scale. This may happen suddenly, affecting important crops and leaving people without food.
environmental changes	This may affect crop growth and reduce food production. For example, if rains fail widespread famine may occur in some countries. Global warming may increase these threats.
the cost of agricultural inputs	Farmers in developing countries where food is scarce may not have the resources to buy fertiliser to maximise crop growth. They may not be able to afford to buy pesticides or disease resistant varieties, so are more likely to lose crops.
war and conflicts in some parts of the world	This may affect the availability of water or food. People may not have access to areas of agricultural land that are disputed or have been made dangerous by war. Conflicts may restrict access to water supply or food imports.

BIOLOGY HIGHER

Main activity: Farming and fishing techniques   Time 15 mins

Learning objectives

- To describe methods that can improve the efficiency of food production
- To understand how application of different fishing techniques promotes recovery of fish stocks

Equipment

none

1. The diagram below shows the layout for two pig farms. One raises pigs outdoors, the other uses intensive indoor methods.

Indoor farmed	Outdoor farmed
pig pens inside a temperature controlled barn	pigs free to move around the field

   a) The indoor farming method is a more efficient method of food production. What does this mean?

   b) Explain why the indoor method shown is more efficient. Discuss at least two reasons.

   c) Pigs and other farm animals are often fed specialised diets to make them grow quickly. What type of food will help to promote fast growth?

   d) Even though the intensive farming method is more efficient, many people think that animals should not be farmed this way. Explain why.

2. Global ocean fish stocks are declining. The graph shows the changes in the biomass of cod of spawning age over a 50 year period (ICES, 2013).

   a) Describe the changes in biomass to your tutor.

   b) Suggest reasons for the changes between 1970 and 2003 and explain why a trend like this is a concern.

   c) Describe two conservation measures that may have resulted in the increase in fish stocks between 2006 and 2013.

   d) Conservation measures aim to maintain fish stocks at a sustainable level. Explain what this means.

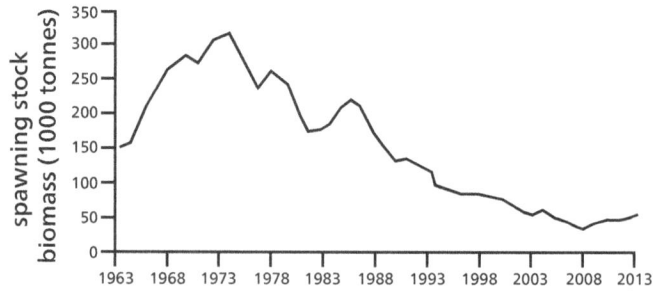

BIOLOGY HIGHER

Main activity: The role of biotechnology

Time: 15 mins

Learning objectives

- To describe and explain some possible biotechnical and agricultural solutions to maintain food security for a growing population

Equipment

none

Modern biotechnology may be able to help meet the food needs of the growing human population. Microorganisms could be cultured for food in large quantities. One such food is made from the fungus *Fusarium*. It is grown to produce a protein-rich food called mycoprotein.

1. Look at the diagram showing how *Fusarium* is grown in a fermenter. Then write down answers to the questions.

    a) Why is glucose syrup added?

    b) Why is it important that air is added?

    c) The ammonia is a nitrogen source. Suggest why this is needed.

    d) Is mycoprotein suitable for vegetarians? Explain your answer.

    e) Suggest why the mycoprotein needs to be purified.

2. Genetically modified bacteria can also be grown in a fermenter to produce human insulin. What is this used for?

3. Genetic modification of crops could be used to improve food security. Draw lines to match up the boxes to show how each modification can help food security.

Genetic modification	How it improves food security
improved nutritional value of crop such as additional vitamins	The crop can be sprayed with herbicide to kill only weeds, reducing crop losses through competition.
improved pest or disease resistance	More food can be produced per area of land.
crop is resistant to herbicide	It may help to prevent malnutrition when people can only grow a small variety of crops.
crop produces bigger fruit or higher yields	It reduces crop losses from damage and from being eaten by herbivores.

242

BIOLOGY HIGHER

Homework activity: Farming efficiency

Time 15 mins

Learning objectives

- To describe some of the biological factors affecting the efficiency of food production

Equipment

none

1. Sara has decided to become a vegetarian. She is explaining why.

> If everyone were vegetarian we could feed the world more easily. You can get much more food from every hectare of land growing cereal than you can from beef.

a) Explain why Sara is correct about the amount of food that can be produced per hectare from cereals compared to beef. Use ideas about food chains in your answer.

_____
_____
_____

Sammy replies with some different ideas.

> You can always use more efficient ways to produce meat. Anyway, it wouldn't be possible to change all farming to only produce plants.

b) Describe two ways in which the efficiency of meat production can be improved.

_____
_____
_____

c) Explain how these methods improve the efficiency of production.

_____

d) Suggest one biological reason why farming to produce only plants might not be practical on a global scale.

_____

38 Answers

Starter activity: Food security

1. The factors are correctly matched on the starter activity sheet before the cards are cut out.

Main activity: Farming and fishing techniques

1. a) A higher percentage of the energy in the food fed to the pigs is transferred into energy in pig biomass. More pork will be produced from every kilogramme of feed supplied.
   b) The indoor method restricts energy transfer from the pigs to the environment. They can't move around much so less energy is wasted in movement. The temperature of the barn is controlled, so they will lose less energy in keeping warm, so more energy from food can go into growth.
   c) Feed that is high in protein
   d) People may have ethical objections to intensive farming. They believe it is cruel to keep animals in pens with no room to move. They believe that animals suffer in these conditions. People may object to the use of other resources, such as burning fuel for heating barns, or using antibiotics to prevent disease in high density pens.
2. a) Cod biomass declined between 1970 and 2006, but after this it shows a slight increase.
   b) Overfishing: more fish removed than could be replaced by breeding; if fish stocks fall too low, breeding will be unable to continue and the species may disappear altogether
   c) Control of net size: don't allow nets above a certain size so fewer fish caught, or set a minimum mesh size so holes are big enough to allow smaller fish to escape and survive to grow and breed; fishing quotas to reduce catches: each government has a quota which it divides between fishing boats which gives the maximum amount of fish that a boat is allowed to catch in a year
   d) A sustainable fishing level is one that can be continued without any future decline in fish stocks.

Main activity: The role of biotechnology

1. a) As an energy source for the fungus
   b) *Fusarium* needs aerobic conditions to grow.
   c) Nitrogen is needed by the fungus to synthesise proteins.
   d) Yes; it is a fungus; no animal products are used to grow it
   e) There may be waste products present, or it may be contaminated with other microorganisms.
2. Harvested and insulin purified; then used to treat people with diabetes
3. Improved nutritional value of crop such as additional vitamins: may help to prevent malnutrition when people can only grow a small variety of crops; improved pest or disease resistance: reduces crop losses from damage and from being eaten by herbivores; crop is resistant to herbicide: crop can be sprayed with herbicide to kill only the weeds, reducing crop losses through competition; crop produces bigger fruit or higher yields: more food can be produced per area of land

Homework activity: Farming efficiency

1. a) Any two points from: energy is lost at each trophic level of a food chain; the food chain is shorter for plants; only two levels rather than three; less energy is lost
   b) Limiting movement of farmed animals; controlling the temperature of their surroundings
   c) Both methods restrict energy transfer from the animals to the environment/reduce energy losses to environment
   d) Not all land/soil is suitable to grow crops; may destroy habitats that rely on grazing/reduce biodiversity; would remove cheap supply of fertiliser/greater use of agrichemicals needed; any other sensible biological suggestion

BIOLOGY HIGHER

Glossary

Abiotic factor
Any non-living component of an ecosystem that affects the population of another organism or its environment

Abundance
The number of organisms per unit area or per unit volume

Accommodation
The process by which the shape of the lens changes to focus on near or distant objects

Active site
The region of an enzyme where substrate molecules bind and undergo a chemical reaction

Adaptation
Acquiring or possessing characteristics (behavioural, physiological or structural) that allow a species to be well suited to and survive in its environment

Amylase
Carbohydrase found in saliva and the small intestine that breaks starch down

Antibody
A specialised immune protein produced by white blood cells (lymphocytes) that bind to specific antigens to fight disease

Antigen
A substance or molecule (usually a protein) that stimulates an immune response, causing the host organism to produce antibodies against it

Antitoxin
A type of antibody with the ability to neutralise a specific biological toxin

Aorta
Artery that carries oxygenated blood under high pressure from the left ventricle to the body

Aseptic technique
Practical steps to prevent contamination with microorganisms from the surroundings or equipment

Basal metabolic rate
The rate of energy expenditure per unit time when at rest. The energy is required for essential body functions such as heartbeat, breathing and maintenance of body temperature.

Benedict's test
A qualitative test for the presence of sugars in a sample (reagent turns from blue to orange in presence of glucose)

Benign tumour
A slow-growing tumour with a defined border that does not spread and is usually harmless

Bile
Alkaline substance produced in the liver and stored in the gall bladder that emulsifies fats to increase their surface area to aid digestion (by lipases)

Binary fission
A form of asexual reproduction in bacteria where a parent cell splits into two independent cells

Biodiversity
The variety of all the different species of organisms on Earth, or within an ecosystem – this is usually includes genetic diversity, as well as species diversity

Biofuel
A fuel composed of or produced from biological raw materials

Biomass
The mass of all living material in a specific area; usually expressed as mass per unit area

Biotic factor
Any living component of an ecosystem that affects the population of another organism or its environment

Biuret test
A qualitative test for the presence of protein in a sample (a blue reagent that turns deep purple in presence of protein)

Carbon sink
Any natural reservoir that absorbs more carbon than it releases as carbon dioxide

Carcinogen
Cancer-causing agent

Cardiac output
A measure of the volume of blood pumped per minute by the heart

Catalyst
A substance that changes the rate of a reaction without being altered itself

Cell function
The job or purpose of a cell

Cell structure
How the cell is organised or put together, including its shape and the sub-structures that it contains

Chlorosis
Loss of the normal green colouration of leaves of plants or yellowing of leaves

Clone
Genetically identical cell or organism

Communicable disease
An illness that can be spread from one person to another

Community
All the animals and plants living and interacting in a particular area

Cone cells
Photoreceptors which are responsible for colour vision and function best in relatively bright light

BIOLOGY HIGHER

Glossary

Coronary arteries
Small blood vessels found on the outside of the heart that supply the heart muscle with oxygen

Coronary heart disease
Cardiovascular condition caused by a build-up of fatty material in the coronary arteries resulting in a reduction in blood flow to the heart muscle

Deamination
The breakdown of amino acids by removal of the nitrogenous part of the molecule

Decomposer
Heterotrophic organism, usually a bacterium or fungus, which breaks down organic material from other dead organisms to form simpler substances

Denature
Irreversible changes to the active site of an enzyme that occur at too high a temperature or too low (too acidic) or too high (too alkaline) a pH

Dendrite
A short branch at the end of a nerve cell, which receives signals from other cells

Differentiation
The process by which a cell becomes specialised in order to perform a specific function

Distribution
The arrangement or pattern of where a species may be found within an area

DNA replication
The process by which a DNA molecule is copied to produce two identical DNA molecules

Double-blind trial
A trial in which neither the patient nor the person carrying out the research knows if the patient is receiving the placebo or the active drug

Double circulatory system
A circulatory system seen in most mammals with two loops, one which links the heart with the lungs and one which links the heart with the rest of the body

Ecosystem
The interaction of a community of living organisms (biotic) with the non-living (abiotic) parts of their environment

Egestion
Process of discharging undigested or waste material that has passed through the gut without being absorbed (defecation)

Embryo
An organism in the early stages of development

Endothermic
Describes a reaction that takes in energy from its surroundings

Enzyme–substrate complex
The temporary complex formed when a substrate molecule binds with the active site of an enzyme

Epidemiology
The study of the incidence, cause and distribution of disease in populations

Eukaryotes
Group of organisms with cells that have genetic material contained in a membrane-bound nucleus and also contain a number of other membrane-bound organelles such as mitochondria and chloroplasts

Evolution
A change in the inherited characteristics of a population over time

Excretion
Removal of metabolic waste products from an organism

Exothermic
(Reaction which) transfers heat energy to the surroundings

Fertilisation
The joining (fusion) of male and female gametes

Fishing quota
A limit on the mass of fish that can be caught annually

Flaccid
A cell is flaccid when the plasma membrane of a plant cell is not pressed tightly against the cell wall. The cell contains less fluid than when turgid but is not plasmolysed (membrane has not fully pulled away from the cell wall).

Gamete
A haploid cell involved in sexual reproduction; two gametes fuse to form the zygote

Gene
A section of DNA which carries the instructions to make a protein; the basic unit of heredity

Genome
The entire genetic material of an organism including both coding and non-coding DNA

Gravitropism/geotropism
A directional growth in response to gravity

Health
A state of physical and mental well-being

Hybridisation
The process of cross breeding which uses different varieties of organisms to create a hybrid offspring with genes from parents of two different varieties

Impulse
Electrical signal (action potential) that passes along a neurone

Glossary

Ingestion
When a substance is taken into the body as food or drink (not necessarily absorbed, may pass through gut)

Inverse square law
(Light) intensity is inversely proportional to the square of the distance from the light source

Iodine
Reagent that turns from brown to blue-black in the presence of starch

Lipases
Group of digestive enzymes that break fats and lipids down into fatty acids and glycerol

Magnification
A measure of how much bigger the image you see is than the real object

Malignant tumour
A fast-growing cancerous tumour that can spread to become secondary tumours in other organs

Meristem
Region of plant tissue, found at the tips of roots and shoots and in the cambium, made up of unspecialised dividing cells which form a variety of new tissue types

Metabolism
The sum of all the reactions in a cell or the body

Mimicry
Where one organism resembles another for protection

Mitosis
A type of cell division that results in two daughter cells each being genetically identical to the parent cell

Monoclonal antibodies
Antibodies that are specific to one binding site on one protein antigen and so are able to target a specific chemical or specific cells in the body. Monoclonal antibodies are produced from a single clone of cells.

MRI (magnetic resonance imaging)
Scanning technique used to capture detailed images of the brain and other organs

MRSA
Methicillin-resistant *Staphylococcus aureus*; a resistant strain of a common type of bacteria often found on the skin, inside the nostrils and throat

Mutation
An alteration of the nucleotide (base) sequence of the genome

Mycoprotein
Food rich in protein produced from fungal cells

Natural selection
The process by which a characteristic of an individual allows it to survive to produce more offspring; variants that have phenotypes better suited to their environment are more likely to survive and pass on their genes to the next generation.

Negative feedback
Mechanisms which return any changes in the internal environment back to a set point

Nephron
The functional unit of the kidney, consisting of a glomerulus and its associated tubule

Non-communicable disease
A condition that is not spread among a population

Obese
A BMI (body mass index) of 30 or more

Optimum pH
The pH at which the rate of an enzyme-controlled reaction is greatest

Optimum temperature
The temperature at which the rate of an enzyme-controlled reaction is greatest

Osmoregulation
The control of water balance in the body

Ovarian follicle
A small fluid-filled sac that contains an immature egg, or oocyte, until it develops and is released

Overweight
A BMI (body mass index) of 25 or more

Ovulation
Release of an egg from the ovary

Pacemaker
Group of cells in the right atrium that controls the natural resting heart rate

Partially permeable membrane
Membrane which allows some small particles (such as water) to pass through it but does not allow other molecules such as sugar/salt to pass through

Pathogen
A disease-causing organism (for example a virus, bacteria, protist or fungi)

Phenotype
The observable characteristics of an organism, caused by a combination of gene expression and environment

Phototropism
A directional growth in response to light

Placebo
A harmless pill or substance given to the patient which does not contain any active drug. It acts as a fake drug.

BIOLOGY HIGHER

Glossary

Plasma
The straw coloured liquid part of blood that carries dissolved substances

Plasmid
A small, circular piece of DNA found in addition to the chromosomal DNA of bacteria

Platelets
Small sticky fragments that help to form blood clots

Population
The number of all the organisms of the same group or species which live in a particular area

Prokaryotes
Group of organisms including bacteria and blue-green algae that have few organelles and do not have the genetic material contained in a membrane-bound nucleus

Proteases
Group of digestive enzymes that break proteins down into amino acids

Protist
Eukaryotic organisms that are not fungi, plants or animals. They may be unicellular or multicellular.

Pulmonary artery
Artery that carries deoxygenated blood from the heart (right ventricle) to the lungs

Pulmonary vein
Vein that carries oxygenated blood from the lungs to the heart (left atrium)

Red blood cells
Biconcave disc-shaped cells that carry oxygen bound to haemoglobin around the body. They have no nucleus and a large surface area to maximise space for carrying oxygen.

Refraction
The bending of a ray of light

Resolution
The shortest distance between two points on a specimen that can still be distinguished as separate points

Rod cells
Photoreceptor cells in the retina that can function in low intensity light; not colour sensitive

Specialised cell
A cell with certain features that allow it to do a particular job in an organism

Speciation
The process in which new genetically distinct species arise from a single original species

Species
A species is a group of organisms that can interbreed to produce fertile offspring

Spore
A reproductive cell that is adapted for dispersion. Does not need to fuse with another cell to grow and develop.

Stem cell
An undifferentiated cell which can give rise to many more cells of the same type, and from which certain other cells can arise by differentiation

STI
Sexually transmitted infection

Stroke volume
The volume of blood pumped by the left ventricle of the heart in one contraction

Therapeutic
Healing, or referring to treatment for a disease

Translocation
The movement of food molecules (sugar, sucrose) through phloem tissue

Transpiration
The process of water movement through a plant and its evaporation from aerial parts, such as leaves

Turgid
Swollen, due to high fluid content; in plants the plasma membrane would be pressed tightly against the cell wall

Variation
Differences in the characteristics of individuals in a population

Vasoconstriction
In response to cold temperatures, blood vessels (arterioles) become narrower, causing a decreased blood flow to capillaries near the surface of the skin to minimise heat radiation from the skin.

Vasodilation
In response to hot temperatures, blood vessels (arterioles) widen, causing increased blood flow to capillaries near the surface of the skin in order to maximise heat radiated from the skin.

Vector
An organism that does not cause disease itself but which spreads disease by carrying pathogens from one host to another

Vena cava
Vein that carries deoxygenated blood back to the heart (right atrium) from the body

Ventilation
The movement of air or water around a gas exchange structure

White blood cells
Cells that fight infections either by producing antibodies (lymphocytes) or by engulfing and destroying foreign pathogens (phagocytes)

BIOLOGY HIGHER

Revise mapping guide

Pearson's *Revise* series provides simple, clear support to students preparing for their GCSE (9–1) exams. Parents may ask you if you know of any independent study resources that they can work through with their child, or you may wish to provide such resources yourself.

We have provided below a mapping guide for each lesson in this pack to a corresponding page in the *Revise* series, to make such recommendations easier for you. See page 5 for a list of recommended titles for students studying AQA GCSE (9–1) Biology.

The Revision Guides and Revision Workbooks for each level correspond page-for-page, so the page references are the same for both.

Lesson		*Revise* AQA GCSE (9–1) Biology Higher Revision Guide Page
1	Diagnostic Lesson	
2	Cell biology: Cell structure and microscopes	Microscopes and magnification **1**; Animal and plant cells **2**; Eukaryotes and prokaryotes **3**
3	Cell biology: Cell specialisation and differentiation	Specialised animal cells **4**; Specialised plant cells **5**
4	Cell biology: Culturing microorganisms	Aseptic techniques **7**; Investigating microbial cultures **8**
5	Cell biology: Cell division	Mitosis **9**; DNA and the genome **78**
6	Cell biology: Transport in cells	Diffusion **11**; Exchange surfaces **12**; Osmosis **13**; Active transport **15**; Extended response – Cell biology **16**
7	Organisation: The digestive system and enzymes	The digestive system **17**; Food testing **18**; Enzymes **19**; Investigating enzymes **20**
8	Organisation: The heart, blood vessels and blood	The blood **21**; Blood vessels **22**; The heart **23**; The lungs **24**; Extended response – Organisation **33**
9	Organisation: Non-communicable diseases	Cardiovascular disease **25**; Health and disease **26**; Lifestyle and disease **27**; Alcohol and smoking **28**
10	Organisation: Plant tissues, organs and systems	The leaf **29**; Transpiration **30**; Investigating transpiration **31**; Translocation **32**
11	Infection and response: Communicable diseases – viral and bacterial diseases	Viral diseases **34**; Bacterial diseases **35**; Antibiotics and painkillers **40**
12	Infection and response: Communicable diseases – fungal and protist diseases	Fungal and protist diseases **36**
13	Infection and response: Defence systems, vaccination, antibiotics & painkillers	Human defence systems **37**; The immune system **38**; Vaccination **39**; Antibiotics and painkillers **40**
14	Infection and response: Drug development and monoclonal antibodies	Antibiotics and painkillers **40**; New medicines **41**; Extended response – Infection and response **45**
15	Infection and response: Plant diseases	Viral diseases **34**; Fungal and protist diseases **36**; Plant disease **43**; Plant defences **44**
16	Bioenergetics: Photosynthesis	Photosynthesis **46**; Extended response – Bioenergetics **52**
17	Bioenergetics: Respiration	Respiration **90**; Responding to exercise **50**; Metabolism **51**
18	Homeostasis and response: Homeostasis and the nervous system	Homeostasis **53**; Neurones **54**; The brain **57**; Reflex actions **55**; Thermoregulation **60**; Extended response – Homeostasis and response **75**

BIOLOGY HIGHER

Revise mapping guide

Revise AQA GCSE (9-1) Biology Higher Revision Guide

Lesson		Page
19	Homeostasis and response: The human brain and eye	Neurones 54; The brain 57; Reflex actions 55; The eye 58; Eye defects 59
20	Homeostasis and response: Hormonal coordination in humans	Neurones 54; The brain 57; Reflex actions 55; Hormones 61; Blood glucose regulation 62; Diabetes 63
21	Homeostasis and response: Water and nitrogen balance	Controlling water balance 64; Kidney treatments 66
22	Homeostasis and response: Human reproduction and contraception	Hormones 61; Reproductive hormones 67; Contraception 69
23	Homeostasis and response: Treating infertility and negative feedback	Hormones 61; Reproductive hormones 67; Contraception 69
24	Homeostasis and response: Plant hormones	Mitosis 9; Meiosis 76; Sexual and asexual reproduction 77; Cloning 88
25	Inheritance, variation and evolution: Types of reproduction and meiosis	Mitosis 9; Meiosis 76; Sexual and asexual reproduction 68; DNA and the genome 78; Cloning 88
26	Inheritance, variation and evolution: DNA structure and the genome	DNA and the genome 78; Genetic terms 80; Genetic crosses 81
27	Inheritance, variation and evolution: Inheritance and sex determination	Variation and evolution 84; Darwin and Lamarck 89; Speciation 90; Evolutionary trees 95; Extended response – Inheritance, variation and evolution 96
28	Inheritance, variation and evolution: Genetic understanding, variation and selective breeding	Variation and evolution 84; Selective breeding 85; Family trees 82; Inheritance 83; Mendel 91
29	Inheritance, variation and evolution: Genetic engineering and cloning	Variation and evolution 84; Selective breeding 85; Speciation 90
30	Inheritance, variation and evolution: Evolution and speciation	Variation and evolution 84; Darwin and Lamarck 89; Speciation 90; Evolutionary trees 95
31	Inheritance, variation and evolution: Evidence for evolution: extinction	Variation and evolution 84; Speciation 90; Extended response – Inheritance, variation and evolution 96
32	Ecology: Classification and communities	Classification 94; Ecosystems 97; Interdependence 98
33	Ecology: Adaptations, abiotic and biotic factors	Adaptation 99; Ecosystems 97; Interdependence 98
34	Ecology: Organisation, cycling of materials and decomposition	Food chains 100; Fieldwork techniques 101; Field investigations 102; Decomposition 104; Cycling materials 103
35	Ecology: Environmental change, biodiversity and waste management	Ecosystems 86; Waste management 95; Deforestation 96; Global warming 97
36	Ecology: Land use, deforestation and global warming	Decomposition 97; Deforestation 108; Global warming 109
37	Ecology: Trophic levels in an ecosystem	Food chains 100; Decomposition 104; Trophic levels 111
38	Ecology: Food production	Food security 112; Farming techniques 113; Sustainable fisheries 114; Biotechnology and food 115

BIOLOGY HIGHER

Progress and observations

Progress and observations

BIOLOGY HIGHER

Progress and observations

 # BIOLOGY HIGHER

Progress and observations

Progress and observations

Progress and observations

Published by Pearson Education Limited, 80 Strand, London, WC2R 0RL.

www.pearsonschools.co.uk

Text © Pearson Education Limited 2018
Series consultant: Margaret Reeve
Edited by Elektra Media Ltd
Designed by Andrew Magee
Typeset by Elektra Media Ltd
Produced by Elektra Media Ltd
Original illustrations © Pearson Education Limited 2018
Illustrated by Elektra Media Ltd
Cover design by Andrew Magee

The right of Janette Gledhill to be identified as author of this work has been asserted by her in accordance with the Copyright, Designs and Patents Act 1988.

First published 2018

21 20 19 18
10 9 8 7 6 5 4 3 2 1

British Library Cataloguing in Publication Data
A catalogue record for this book is available from the British Library

ISBN 9781292201412

Copyright notice
All rights reserved. The material in this publication is copyright. Activity sheets may be freely photocopied for use by the purchasing tutor. However, this material is copyright and under no circumstances may copies be offered for sale. If you wish to use the material in any way other than that specified you must apply in writing to the publishers.

Printed in the United Kingdom by Ashford Colour Press Ltd

Acknowledgements
We would like to thank Tutora for its invaluable help in the development and trialling of this course.

Notes from the publisher
Pearson has robust editorial processes, including answer and fact checks, to ensure the accuracy of the content in this publication, and every effort is made to ensure this publication is free of errors. We are, however, only human, and occasionally errors do occur. Pearson is not liable for any misunderstandings that arise as a result of errors in this publication, but it is our priority to ensure that the content is accurate. If you spot an error, please do contact us at resourcescorrections@pearson.com so we can make sure it is corrected.